The Oil Kingdom at 100

Petroleum Policymaking
in Saudi Arabia

Nawaf E. Obaid

Policy Papers no. 55

THE WASHINGTON INSTITUTE FOR NEAR EAST POLICY

© 2000 by the Washington Institute for Near East Policy

Published in 2000 in the United States of America by the Washington Institute for Near East Policy, 1828 L Street NW, Suite 1050, Washington, DC 20036.

Library of Congress Cataloging-in-Publication Data

Obaid, Nawaf E., 1974–
    The oil kingdom at 100 : petroleum policymaking in Saudi Arabia / Nawaf E. Obaid.
        p.   cm. — (Policy papers ; no. 55)
    Includes bibliographical references.
    ISBN 0-944029-39-6 (pbk.)
    1. Petroleum industry and trade—Government policy—Saudi Arabia. I. Title: Oil kingdom at one hundred. II. Title: [Sub-title variant:] Petroleum policy making in Saudi Arabia. III. Title. IV. Policy papers (Washington Institute for Near East Policy) ; no. 55.
    HD9576.S33 O22 2000
    338.2'7282'09538—dc21   00-022055
                                                           CIP

Cover photo © Larry Lee/Corbis.
Cover design by Monica Neal Hertzman.

The Author

Nawaf E. Obaid graduated from Harvard University's John F. Kennedy School of Government with a master's degree in public policy. He completed his undergraduate studies at Georgetown University's Edmund A. Walsh School of Foreign Service, receiving a bachelor of science degree in foreign service. He is from Medina, Saudi Arabia.

Obaid was a 1999 visiting research fellow at The Washington Institute for Near East Policy.

• • •

Table of Contents

Acknowledgements

I would like first to thank my parents—my mother for her unwavering moral and emotional help and my father for his crucial financial support—for all the encouragement they have shown me, especially in the most difficult moments of this project. I am also eternally grateful to my two brothers, Tarek and Karim, who never questioned my decision to undertake this project and always stuck by me, even at the worst of times. Also, many thanks go out to my Saudi cousins, Fadi Z. Ghazzawi, Sultan O. Ghazzawi, and Amr Z. Ghazzawi, for their hospitality while I was working on the project. And last but not least, Tamara, my Jordanian fiancée, who put up with me and saw this project through to the end.

My sincere greetings also go out to the senior research staff of The Washington Institute. I would like to thank Patrick Clawson, the Institute's director for research, for all the time and effort he devoted to working with me, sharing ideas, and discussing issues. Special thanks also go out to Michael Eisenstadt, Alan Makovsky, Ray Takeyh, and David Schenker for all the great laughs we had together. On a more serious note, I would also like to particularly thank Robert Satloff, the Institute's executive director, for having had me at the Institute to do such a project.

I would also like to thank all the outside readers of the book who really gave me invaluable comments. My thanks go out to Richard Dekmajian of the University of Southern California; former U.S. ambassador to Saudi Arabia Walter Cutler, president of Meridian International Center; Ed Morse, former publisher and president of the Energy Intelligence Group; Nathaniel Kern, president of Foreign Reports Inc.; and last but not least, Joseph Nye, dean of the Kennedy School of Government at Harvard University.

My thanks also go out to Chad Gracia, president of Gracia Group of Companies, a New York-based research consultancy, for gathering all the background material and notes for this project. His work saved me invaluable research time.

Finally, I would also like to mention the Institute's administrative staff for all the help they gave me. My thoughts go out to Monica Neal Hertzman, the Institute's director of publications; Alison Rogers, the executive assistant to Dr. Satloff; Carmen Perez, the Institute's former financial assistant; and 1998–99 research assistants Adam Frey, Ben Orbach, and Assaf Moghadam.

Nawaf E. Obaid
Saudi Arabia
February 2000

Preface

Politically and militarily, Saudi Arabia plays a central role in the stability of the difficult Persian Gulf neighborhood. While rogue regimes in Iraq and Iran pursue weapons of mass destruction and aim to undermine the Middle East peace process, U.S. military forces have been deployed to Saudi Arabia for a decade now, defending against these threats. Economically, Saudi Arabia is also important. Oil prices are now at their highest level in a decade, and with one-fourth of the world's oil reserves, Saudi Arabia is the world's largest oil exporter. The United States, by far the world's largest petroleum importer, would do well to understand how the kingdom makes its petroleum policies and influences international production.

Yet Saudi Arabia is not an easy society to understand. How the kingdom functions can appear murky even to the informed outsider. We are therefore pleased to present this detailed analysis of Saudi petroleum policymaking by Nawaf Obaid, a young Saudi graduate of the Kennedy School of Government at Harvard University who was a visiting fellow at The Washington Institute. Obaid argues that decision making about petroleum is becoming more formal, professional, and bureaucratic, rather than based on personal influence and whims. Obaid documents the Saudi desire to open up more to market forces and foreign investment—an interest brought out most clearly in the debates about what role international oil companies should play in future gas, oil refining, and oil production projects. He argues that petroleum provides a lens through which Saudi policy as a whole can be better understood.

The trends Obaid identifies are a mixed blessing from the perspective of U.S. interests. Differences have always existed concerning Israel and, at various times, about how to deal with the difficult regimes in Iran and Iraq. More assertive and nationalist Saudi policies may well run counter to U.S. policies. On the other hand, it is also likely that better Saudi governance will make for a more prosperous and stable kingdom, effectively eliminating the risk of political unrest that could bring to power radical Islamists of the Usama Bin Ladin ilk. The increasingly technocratic and professional leadership that Obaid describes is, moreover, a system of government with which Americans are more comfortable than they are with autocratic rule. A more open Saudi Arabia, more comfortable with foreign investment in crucial fields like petroleum and electricity, is better not only for American business, but also for people-to-people relations between two quite different cultures.

We present this study to better inform policymaking about Saudi Arabia, not to applaud or deplore changes underway in the kingdom. Too much concern is heard about an Iranian-style Islamist revolution or an overthrow of the Saudi ruling family, and too little attention is directed to the changes in how the Saudi government actually works. The Washington Institute is particularly proud to be able to publish a study by this highly regarded Saudi scholar. That a Saudi would author such a detailed, scholarly analysis of his country's petroleum policymaking is emblematic of the increasing professionalism Obaid describes.

Michael Stein   Fred S. Lafer
Chairman   President

Executive Summary

With its 260 billion barrels of proven petroleum reserves, Saudi Arabia is one of the world's key economies and the largest supplier of oil. Its 1999 gross domestic product (GDP) was $140.5 billion, 30 percent larger than any other Middle Eastern country.

Many Western appraisals of Saudi Arabia have argued that the kingdom is the "world's largest family business," and a poorly run one at that. The stereotype of the corrupt and incompetent Saudi prince controlling the levers of power is, however, an inaccurate characterization. Saudi Arabia is rapidly modernizing, particularly in the oil sector. As opposed to policymaking based on personal influence and intuition, Saudi Arabia is moving toward a more formal, professional, and bureaucratized system—one that more closely resembles the Western model of public administration. To be sure, change comes more slowly than foreign observers might like, because Saudi decision making is a cautious and consensus-building process—but as the historical records shows, this is an approach that has served Saudi Arabia well.

Major Decision Makers

Understanding Saudi oil policy requires examining the main actors responsible for policy formation and implementation. Although these individuals hold their positions by virtue of their birth, their records show they have maintained peace and stability in the kingdom during their tenure, when so many others predicted doom.

King Fahd ascended to the throne in June 1982 and named a half-brother, Prince Abdullah, the crown prince. After a stroke in 1995, the Saudi monarch transferred most of the daily responsibilities of running the kingdom to the crown prince, while the king maintains a symbolic role.

Crown Prince Abdullah was until recently called anti-American by many in the Western press. The crown prince does place his country's strategic interests first and has differed from Washington in areas such as the speed of his country's recognition of Israel. At the same time, he has encouraged strong relations with the West; indeed he is widely respected in Saudi Arabia for being able to maintain close ties to the United States without being viewed as subservient to Washington. Overall, he has developed a reputation as an active and popular leader who has taken some bold moves, including calling for the average Saudi to become less reliant on the government. He has made increasing use of those with technical expertise and professional experience, as illustrated by the active role of professionals in analyzing proposals submitted by Western oil companies wishing to invest in the kingdom's petroleum sector. In this way, the policymaking process is moving closer to that of any well-run bureaucratic organization or industrialized nation.

Prince Sultan, the second full brother of the king, is likely the next in line for the throne after Crown Prince Abdullah. Known as the father of the modern Saudi armed forces, Sultan has overseen the military's development and modernization since his appointment as minister of defense in 1962. In this capacity, he oversees a quarter of the Saudi budget. While Western observers admit that the Royal Saudi Land Forces and the Royal Saudi Navy still have to progress enormously, the Royal Saudi Air Force has become the most modern and powerful Arab air force.

The Saudi Petroleum Community

The petroleum sector provides a lens to get a clearer picture of the changes occurring throughout the Saudi government. Because oil policy has such an enormous affect on the health of the kingdom, it is set not by the whim of any individual but instead by consensus among the influential ruling family members after considerable debate and consultation with Saudi experts. For example, in mid-1999, the kingdom adopted a strategy to maintain per-barrel prices in the $18-$20 range for the Brent benchmark crude oil. Some argued that high prices should be maintained at all costs, mainly by cutting production, while others argued that achieving market share—by increasing production and lowering prices—would be the best strategy. These differences should not be exaggerated. None of the leading players in the Saudi petroleum community are extremist proponents of either strategy; the differences are over achieving the right "blend" of policy over time.

Many individuals at many levels participate in what can be termed the Saudi petroleum community. This study examines three main organizations—the Supreme Council for Petroleum and Minerals Affairs, the Ministry of Petroleum and Mineral Resources, and Saudi Aramco—as well as two ad hoc committees that were dedicated to evaluating foreign investment in the kingdom's energy sector. Detailed information is presented about the key actors in the Supreme Council, namely, Minister of Foreign Affairs Prince Saud, Minister of Petroleum and Mineral Resources Ali Al Naimi, Minister of Finance and National Economy Dr. Ibrahim Al Assaf, Minister of Industry and Electricity Dr. Hashim Yamani, Minister of Planning Khaled Al Ghosaibi, President of King Abdulaziz City for Science and Technology Dr. Saleh Al Adhel, former Deputy Minister of Finance and National Economy Abdulaziz Al Rashid, and Saudi Aramco President and CEO Abdallah Jumah. Similar information is presented about the key actors in the ad hoc committees—Minister of State Dr. Mohammed Al Sheikh, Minister of State Dr. Musaed Al Ayban, Deputy Minister of Oil Prince Abdulaziz bin Salman, and Deputy Foreign Minister Dr. Youssef Al Sadoun. The backgrounds of Prince Faisal bin Turki, a senior adviser in the Oil Ministry, and Sulaiman Al Herbish, Saudi Arabia's governor to OPEC, are also presented in detail.

Responses to Weak Oil Prices

The 30 percent drop in petroleum prices during 1998 was disastrous for the Saudi treasury. By the close of 1998, as the real price of oil dipped to 1973 levels, Saudi Arabia entered its first recession in six years. The government reacted by (1) instituting domestic austerity measures, (2) orchestrating petroleum production cuts to raise oil prices, and (3) beginning the process of bringing international oil companies (IOCs) back to the kingdom's energy sector. In each of these moves, the government achieved its policy goals. The combined impact of the austerity measures and the higher oil prices are forecast to reduce the 1999 government budget deficit by $6 billion, equivalent to 11 percent of GDP.

Austerity Measures. To deal with this sharp decline in revenue, the crown prince acted quickly with a series of measures to cut expenses. All government ministries were ordered to cut spending by 10 percent, recruitment and wages in the government sector were frozen, and payments of bills were extended to the maximum time period allowed in the contracts with suppliers. The 1999 budget continued the austerity measures and revealed a general willingness by the Saudi government to adapt its development prerogatives to the economic realities imposed by sinking oil prices. The budget for social subsidies was reduced 50 percent from its 1998 levels; the defense budget, by about 25 percent (exact data are not published). The Saudis also raised taxes and fees, including a tax on luxury goods, higher prices for gasoline (now $0.99 a gallon), an increase on the permit fee for foreigners working in the kingdom, and an airport tax. The increased taxes and fees are expected to raise revenue equal to about 1 percent of GDP; that is the equivalent of a $80 billion a year tax increase in the United States.

Production Cuts. In a series of high-level meetings that preceded the March 1999 meeting of the Organization of Petroleum Exporting Countries (OPEC), the Saudis agreed to renegotiate Iran's quota, and they stunned many analysts by offering to slash their own production by 585,000 barrels per day (bpd). At the meeting, Saudi-orchestrated production cuts were adopted, and prices quickly rose to what Saudis view as the "sweet spot" of $17–$20 per barrel. Within a month of the OPEC meeting, 40 percent of the 1998 price decline was recovered. By summer, prices had recovered 90 percent of the previous year's drop. From the Saudi perspective, orchestrating of the production cuts—and the resulting hike in prices—was a success. It led to a much-needed revenue hike, as what was gained from higher prices more than made up for lower production. Moreover, the Saudis were able to reinforce OPEC solidarity and provide a means for a breakthrough in relations with Iran while at the same time securing acquiescence from the oil consuming nations, chief among them the United States.

Bringing Back U.S. and Other Foreign Oil Companies. In September 1998, seven executives of U.S. oil companies met with Crown Prince Abdullah and Foreign Minister Prince Saud at the Washington residence of Saudi Ambassador Prince Bandar Bin Sultan. At this gathering, the executives were asked to prepare proposals for "mutually beneficial" petroleum projects in the kingdom. IOCs, both American and European, submitted various proposals focusing on oil and gas production ("upstream") rather than on refining and processing ("downstream"). Senior officials in the oil ministry and Saudi Aramco were cool toward the IOC bids, however, arguing that Aramco could perform as well as the IOCs and that as long as Saudi Arabia has substantial excess production capacity, there is little reason to invest in upstream oil projects. Even if the upstream oil sector remains for some years closed to foreign investment, the prospect that it may open up increases the interest of American companies in expanding their presence in the kingdom by taking advantage of other investment opportunities in Saudi Arabia, such as in the petrochemical industry, power plants, and desalination plants. Attracting investments to such sectors seems to be the main goal of the Saudi leadership, and they are on track to achieve this without having to deal with the sensitive topic of opening up the kingdom's upstream oil sector. It is worth noting that there is one area where foreign firms have long produced oil, namely the Saudi section of the "Divided Zone" on the Kuwaiti border. The concession, held to date by a Japanese firm, is up for renewal, and a U.S. firm could well be offered the opportunity.

Reducing Dependence on Oil Income

Despite the improvements that accompanied the rebounding oil prices, the Saudi economy still suffers from major structural problems. Massive subsidies, inefficient industries, and the lack of a fully developed free-market system all hamper the ability of the state to pull itself out of debt. As Crown Prince Abdullah said in a widely noted 1996 speech, the era of the oil boom is over. The Saudi leadership is becoming increasingly wary of relying on a single product (petroleum) for such a large portion of state income and wishes to stimulate a diversification of the economy. The budget problems and the desire for diversification work together to make strengthening the private sector a particularly appealing policy.

Few in Saudi Arabia see the same urgency for privatization as do foreign economists; most Saudi policymakers support privatization in principle but have real qualms about each specific case. Saudi Arabia is proceeding at the same pace as did other reforming states in the Middle East, all of which talked for years about privatization without doing much. Indeed, privatization has been discussed in the kingdom since the Fourth Development Plan of 1985-1990. The kingdom's telephone system is being "corporatized" to prepare it to be the first privatization of a government utility—a step motivated by strains on the telecommunications network requiring investment that the government is ill-placed to make. The power sector is perhaps the most important part of the economy that the government is intent on privatizing. But full privatization of the power companies would force the government to implement some difficult policies. Subsidies must ultimately be removed—but this move is not politically appealing. If they are not removed, then the government will have to make up the difference between the market price and the subsidized price of electricity.

Relations with Other Oil Producers

Saudi foreign policy is in some ways a subset of its oil policy. As such, it has two distinct branches: its relations with fellow exporters and relations with the oil consuming countries of the world. Saudi Arabia's relations with other oil exporters are defined by one key set of facts: The kingdom possesses the largest petroleum reserves, the greatest spare production capacity, and the cheapest uplift costs in the world. Because of this reality, the kingdom has the ability to flood the market, drive out most other high- and medium-cost producers, and ultimately become the world's preeminent oil supplier.

Iran. Saudi Arabia and Iran had hostile relations—fed by mistrust and stereotypes between Sunnis and Shiites as well as between Arabs and Persians—for nearly two decades after the 1979 Iranian revolution. In 1998, however, the two countries began to transform their relationship from an adversarial to a cooperative one, a shift cemented by agreement over the March 1999 oil production cuts which led to sharply higher revenue for both countries. In 1999, the Saudi minister of defense visited Tehran, and Iranian president Muhammad Khatami visited Riyadh. The United States publicly blessed this rapprochement, although many believed that the administration had reservations about the speed with which it occurred.

Iraq. The Saudis have been one of the main beneficiaries of the post-1990 restrictions on Iraq's oil exports. Those restrictions have allowed the other oil producers, including Saudi Arabia, to sell more oil than they could have otherwise. Therefore, Saudi Arabia must watch closely what is happening to the sanctions on Iraq to ensure that any increases in Iraqi production do not result in sharp downward pressures on prices. Despite the fact that the Saudis have a strong financial interest in keeping Iraqi oil off the market, nevertheless Crown Prince Abdullah in January 1999 put Saudi Arabia's full weight behind his proposal to remove all restrictions on the sale of Iraqi oil; his priority was to reduce the humanitarian problems of the Iraqi people. The crown prince proposed simultaneously to retain tight control on what Iraq can purchase in order to pressure Iraq to comply with the UN weapons control requirements.

Gulf Cooperation Council (GCC). Saudi relations with the

five other members of the GCC are greatly affected by issues other than oil. Traditionally, Saudi Arabia has close ties with three (Bahrain, the United Arab Emirates, and Kuwait) and colder relations with two (Oman and Qatar). Relations with Bahrain are closest: Bahrain's well-being rests on economic and security guarantees by Saudi Arabia, as it has limited natural resources and a serious problem in relations between the Al-Khalifa ruling family and the Shi'i majority population. Kuwait and Saudi Arabia have long had a strong relationship, founded on Kuwait's security needs; in OPEC, Kuwait almost always follows the Saudi lead, saving its criticism of Saudi oil policy for private surroundings. In the United Arab Emirates (UAE), the ruling families of Abu Dhabi and Dubai have generally deferred to the kingdom on matters of security and foreign policy, though tensions have started to rise in the past few years, especially over the Saudi development of the large Shaybah oil field on the border between the states. The relationship between Qatar and Saudi Arabia has been strained; issues include the border—Qataris have never been happy with the 1974 UAE-Saudi border agreement which they saw as being at their expense—and the Qatari television station Al-Jazeera (founded in 1996), which tackles controversial issues head-on. Relations between Oman and Saudi Arabia have always been distant and cold: Sultan Qaboos is an Ibadi, a movement which, in the view of Saudis, is only loosely connected to Islam.

Venezuela and Mexico. The Saudis have long been suspicious of the Venezuelans, accusing them of cheating on production agreements in order to increase their revenues and solidify their position in the U.S. market. Yet, the drastic price drop of 1998 forced the Saudis to overlook these grudges and try to find common ground. In March 1998, Prince Abdulaziz bin Salman, the deputy oil minister for petroleum affairs, was approached by non-OPEC Mexico, and the prince then orchestrated a meeting in Riyadh and one in Amsterdam between the three producers. According to Mexican Energy Secretary Luis Tellez, Mexico agreed to "act as a middleman" between the Saudis and the Venezuelans. The result was a production agreement between the three main exporters to the U.S. market. Although these two meetings were unable to staunch the drop in prices, they laid the groundwork for the ultimately successful Hague meeting in March 1999.

Once again, it is clear that the demands of the Saudi oil policy—a need to shore up prices—led to rapprochement between former disputants. In the future, relations between the kingdom and these two countries can be expected to remain a priority, by virtue of the fact that they are the three main suppliers to the United States.

Relations with Oil-Consuming Nations

Saudi Arabia has been working in recent years to improve its relations with the world's main oil consuming nations. It has engaged in proactive marketing and sales arrangements, as well as invested in petroleum projects with consuming nations. Such an active marketing policy is an indication that Saudi Arabia intends to maintain oil prices at the $18–$20 level, a price at which so much production comes on to the market that producers must struggle to identify markets for their output and must work to protect those markets from others. If Saudi Arabia intended to flood the market with cheap oil, it would worry less about the competition because it could rest confident that its low production cost would guarantee that it would come out on top.

Asia. Asia took 60 percent of Saudi Arabia's crude exports before the 1997 Asian financial crisis. The Saudis realize that in the long term, its leading customer is likely to be China, whose oil consumption has been rising fast. The Saudis have worked to consolidate their presence in the Asian market by establishing joint ventures with Asian companies, such as South Korean and Philippine refineries.

Europe. Over the past decade, Saudis have made several key purchases in European oil companies, including Greek and Swedish refineries. But Europe is becoming less important to them as an export destination because of its high taxes that limit demand for oil, its alternative sources of petroleum from closer-in (Libya, the North Sea), and its strong ties to other oil exporters (Iraq, Algeria).

The United States. Oil is one of the two pillars of the U.S.-Saudi "special relationship," the other being security. The United States, which consumes 25 percent of the world's average daily output of oil, imports more than half of its oil. Saudi Arabia is determined to remain a large supplier to the U.S. market; in most years, it is the single largest, just ahead of Venezuela. The United States is a vital market for Saudi Arabia, accounting for 22 percent of all its exports. Saudi Aramco is partners with Texaco and Shell in Motiva Enterprises, which owns thirteen refineries among its $13 billion in assets. With the invitation to U.S. oil companies to reenter the kingdom's energy sector and the general opening of the Saudi economy, the two countries' economic interrelations seem poised to grow.

The strategic alliance has been one of America's most enduring. For decades, U.S. administrations have made strong ties to the kingdom one of America's foreign policy priorities, even though the two do not share a formal defense treaty. The strategic relationship has rarely been strained (the obvious exception being the 1974 oil embargo), though there are disagreements—most strongly about issues relating to Israel, as well as about proliferation of nonconventional weapons and the use of force against Iraq.

Conclusions and Recommendations

The oil sector illustrates trends throughout Saudi policy. The crown prince and the senior Saudi leadership have adopted an increasingly professional approach to policymaking and implementation. They are open to a larger role for foreign investment and towards structural economic reform. They are also taking a more assertive stand in regional and international politics. In short, Saudi policy is becoming more professional and more assertive.

U.S.–Saudi Relations. Some elements in the U.S. policymaking establishment may perceive Saudi Arabia's emerging assertiveness and independence in foreign affairs as a threat to the partnership. An important fact to keep in mind is that if the Saudi monarchy were seriously undermined, the delicate power sharing between the royal family and the religious establishment would be unbalanced. Despite drastic differences in religion, culture, and politics, the United States and Saudi Arabia have found a way—inspired by mutual need and shared strategic goals—to coexist in respectful friendship. Consensus on all issues is impossible; differences will persist, especially those involving Israel and Muslim-majority states hostile to the United States (e.g. Iraq and Iran).

Reorganizing OPEC. OPEC's membership is irrational, excluding such large oil exporters as Mexico while including such small exporters as Qatar. The interests of the oil producers would be better served if OPEC were reinvented or replaced by an "OLPEC"—an Organization of Large Petroleum Exporting Countries. Such a grouping would have enormous power to manage oil production and prices. That would have a mixed impact on the United States: as a large oil consumer, it benefits from low prices, but the large, high-cost U.S. oil industry benefits from high prices, and the entire U.S. economy benefits from more stable prices.

Saudi Oil Policy. Should the kingdom work to maintain prices through global production management, or should it instead focus on increasing its market share, regardless of price? A policy of high production levels would drive prices down. That would have several advantages over the long term. Lower prices would put high cost competitors such as Britain and Norway out of business in the long term, and would spell the end of new oil development such as that planned in Central Asia. It would also slow the ultimate adoption of emerging non-oil energy technologies such as the electric car or the hydrogen fuel cell. But the disadvantage of a high-production, low-price policy would be that, in the early years, Saudi Arabia would lose tens of billions of dollars in revenue, because prices would fall immediately while production levels would increase only over time. To prevent severe economic dislocations and social unrest, the Saudis would have to raise substantial funds to cover the gap. One way to mobilize that revenue would be to move forward with the long planned sell-off of state enterprises.

Chapter 1

Introduction

Saudi Arabia, with 260 billion barrels of proven petroleum reserves, is one of the world's most important economies and its largest supplier of oil. Its 1999 gross domestic product (GDP) was $140.5 billion, 30 percent larger than any other Middle Eastern state.[1] Because of Saudi Arabia's substantial contributions to the World Bank and the International Monetary Fund (IMF), it appoints permanent executive directors at both institutions, just as do Japan, Germany, and the permanent United Nations Security Council members. In addition to Saudi Arabia's economic assets, the kingdom is custodian of Islam's two holiest cities—Mecca and Medina—further ensuring that Saudi Arabia will remain a key actor in the Islamic and Arabic worlds. For all these reasons, Saudi Arabia is vitally important to the United States.

Yet, despite the kingdom's strategic significance, Western appraisals of Saudi Arabia have long been characterized by ignorance at best and condescension at worst. Many Western observers perceive the Saudis as merely a corrupt and inefficient dynasty and their economic decisions as nepotistic and shortsighted more than anything else. In 1982, one magazine noted that most Saudis work "with awesome incompetence,"[2] an opinion shared by many even today. Many business and government leaders perceive Saudi Arabia as the "world's largest family business" and a poorly managed one at that. In March 1999, analysts wrote that policymaking in the kingdom has been stymied by "a general paralysis in decision making, which reflects the ambiguous state of the top leadership."[3]

Although usually making a more favorable assessment of the Saudi leadership, U.S. government analyses have long misunderstood the kingdom. In fact, those who monitor Saudi affairs most closely admit the country baffles them. Ronald Neumann, U.S. deputy assistant secretary of state for Near Eastern affairs, admits, "We don't understand how the Saudis make decisions."[4] For example, a secret, internal Central Intelligence Agency (CIA) memo, "Short Term Outlook in Saudi Arabia," dated October 19, 1962, deeply underestimated the fortitude of the ruling family. It concluded that "if [King] Faisal is able to ride out the Yemeni crisis, we believe he will be able to preserve royal rule for at least the next year or so."[5] It also contained incorrect statistics, ignored the influence of the senior princes, and failed to accurately identify the true decision makers in the kingdom. Neither did it consider social and economic constraints imposed on the leadership, regional considerations, or the unique position of the kingdom in the Islamic world. Such omissions have been common in official analyses since, and rectifying this general deficiency is one of the main purposes of this book.

At the dawn of the twenty-first century, Saudi Arabia is developing the administrative apparatus of a modern state with a well-trained bureaucracy and strong leadership. The stereotype of the corrupt and incompetent Saudi prince controlling the levers of power is simply not correct. Certainly, change comes slowly, because Saudi decision making is cautious and based on consensus-building—an approach that has served the kingdom well. Despite problematic episodes in its history (the Free Princes episode of the early 1960s and the 1979 Qa'aba takeover by Islamist fundamentalists), no other major country in the region can claim Saudi Arabia's record in avoiding revolution, war, or social upheaval.

Although change may seem gradual from year to year, great transitions have characterized Saudi Arabia for much of this century. In the 1920s, King Abdulaziz (better known as Ibn Saud) transformed the peninsula from a collection of desert tribes into a unified state. In the 1970s and 1980s, massive oil wealth changed the kingdom from an undeveloped backwater to one of the world's richest countries. Now,

Saudi Arabia: Key Facts (1999)	
Area	2,150,000 sq. km.
Political System	Absolute Divine Monarchy
Head of state	King Fahd
Population*	
(Saudi)	7 million–8 million
(Expatriate workers)	3 million–4 million
Gross Domestic Product	
(GDP) in millions	$140,500
GDP per capita**	$12,772
Oil reserves	261.2 billion barrels
—share of world	25.9 percent
Gas reserves	185.9 trillion cubic feet
—share of world	3.7 percent
Total imports	$38.8 billion
Total exports	$54.5 billion

Notes: *The official population count is a highly questionable 20 million. The breakdown puts total Saudi citizens at 14 million, with expatriates at 6 million.
**The official figure, based on inflated population estimates, is $7,200.
Sources: *World Development Indicators 1999*, *World Bank Group Directory of Trade Statistics 1999*, International Monetary Fund estimates, and confidential personal estimates based on interviews with U.S. government agencies.

at the end of the 1990s, Saudi Arabia is moving even further along the road of modernization, especially in the oil sector, where the kingdom's oil policy is growing more professional, liberal, and assertive.

The next chapter will provide an overview of the main Saudi decision makers with authority over oil policy and will examine their experiences, qualifications, and views. For insight into how the kingdom may conduct a partial liberalization of the oil industry incorporating private and foreign participation, the following chapters will study how Saudi decision makers are preparing the kingdom's domes-

tic economy for the next century, will closely analyze the petroleum sector, and will briefly explore efforts to privatize
other sectors of the economy, including electric utilities and
the national Saudi airline. Finally, this book will analyze how
Saudi Arabia's oil policy affects its regional and international
relations, both with fellow exporters and with oil-consuming
countries.

Notes

1. Throughout this book, the term "Middle Eastern states" will refer to
   all Arab League member states, Iran, and Israel, but not Turkey. Information on the Saudi GDP is derived from the *World Bank Yearly
   Report 1999*, the Saudi Ministry of Finance, and independent estimates.

2. "Filling the Void: Eye of the Needle," *Economist*, February 13, 1982,
   p. 4.

3. Robin Allen et al. "Running on Empty," *Financial Times*, March 29,
   1999, p. 15.

4. Author interview in Washington, November 1997.

5. Sherman Kent, *The Board of National Estimates*, obtained by the author
   through a Freedom of Information Act request, Mori Doc ID: 14220,
   p. 1.

Chapter 2

The Senior Leadership
of the Royal Family

For the past half-century, oil policy in Saudi Arabia has been made as all other policy is made in the kingdom—in an ad hoc, informal manner. For the most part, the Saudi king has conducted the kingdom's foreign and domestic policy, a method characteristic of the general nature of decision making in the Arab monarchies of the Persian Gulf. On many issues, this approach maintains a positive record of success; for example, from 1974 to 1994, the kingdom's social, health, and infrastructure indicators rose more quickly than almost any other developing country.[1]

At the beginning of the twenty-first century, however, a new model is emerging. Instead of policymaking based on personal influence and intuition, Saudi Arabia is moving toward a more formal, professional, and bureaucratized system that closely resembles the Western model of governance. Crown Prince Abdullah, who "wants to turn Saudi Arabia into a modern nation able to compete in the next century," is leading this transformation.[2] Although this spirit of change has pervaded every important Saudi government institution, it is most evident in the petroleum sector.

Understanding Saudi oil policy requires an examination of the main actors responsible for policy formation and implementation. Although these individuals hold their positions by virtue of birth, an analysis of their records shows they have generally been efficient managers and leaders, each of whom has focused on a specific area in the kingdom's development and has gained wide experience in that sector through more than two decades of public service. For example, the crown

5

prince and the minister and vice minister of interior have concentrated on the political and social stability of the kingdom, the minister of defense has overseen the expansion and modernization of the Saudi armed forces, and the governor of Riyadh has been the main force behind both the capital city's huge growth and the effort to manage the urbanization of native Saudis. Interestingly, they have all monitored petroleum matters, the acknowledged backbone of the domestic economy. Many have begun preparing for the future by ensuring that their sons are qualified to lead the country when the time comes. They recognize that the younger generation of princes will provide the core of future leadership in line with Saudi traditions that emphasize the importance of genealogy as well as merit. Nonetheless, in the final analysis, the best proof of the fitness of the current leaders is that they have maintained peace and stability in the kingdom during their tenure, when so many predicted doom.[3]

King Fahd
Custodian of the Two Holy Mosques; Prime Minister

Then-Prince Fahd was the kingdom's first minister of education from 1954 to 1960. In 1962, he was named minister of interior, a post he held until 1975. In that year, the newly crowned King Khaled named him crown prince. For the next seven years, the crown prince oversaw the day-to-day Saudi government administration, including the massive infrastructure development of the late 1970s and 1980s. Upon his accession to the throne on June 13, 1982, King Fahd named as crown prince one of his half-brothers, Abdullah.

In 1990, faced with a potential invasion by Iraq, King Fahd invited U.S. troops into Saudi Arabia. This move met strong domestic and regional opposition but succeeded in protecting the kingdom and liberating Kuwait. After a stroke in 1995 and the removal of his gall bladder in 1998, the Saudi monarch lost much of his strength. By the close of the century, King Fahd had transferred most of the daily responsibilities of running the kingdom to Crown Prince Abdullah, who has acted as de facto regent since January 1, 1996. Although the king maintains a symbolic role, by the closing years of the decade he chose to exercise very little sway over Saudi policy.

The Next Generation

The Saudi monarch has six sons. His eldest, Prince Faisal, was president of the Youth and Welfare Organization, overseeing all sporting activities in the kingdom before his death in August 1999. King Fahd's second son, Prince Mohammad, is governor of the Eastern Province. Prince Saud is the deputy director of the General Intelligence Directorate. Prince Sultan is the vice president of the Youth and Welfare Organization. Prince Khaled is a private businessman. The Saudi monarch's youngest son, Prince Abdulaziz, is a minister of state without portfolio.

Sons of King Fahd Who Hold Official Positions	
Prince	**Position**
Mohammad	Governor, Eastern Province
Saud	Deputy Director, General Intelligence Directorate
Sultan	Vice President, Youth and Welfare Organization
Khaled	Private Businessman
Abdulaziz	Minister of State without Portfolio; Chief of the Council of Ministers

Crown Prince Abdullah
Vice Prime Minister; Commander of SANG

Crown Prince Abdullah, the son of King Abdulaziz and half-brother to King Fahd, was born in 1928;[4] he was designated heir to the throne in 1982, and has also been commander of the Saudi Arabian National Guard (SANG) since 1963. The crown prince was acting king for a month in 1995 after King Fahd suffered a stroke, and since 1998 he has essentially assumed management of the day-to-day operations of the government. During this time, he has developed a reputation as an active and popular leader. One British journalist noted recently that the crown prince "conveys warmth and simplicity, talks to ordinary people, tells home truths. Above all, no odour of scandal attaches to him."[5] Another applauded that "he has talked honestly to Saudis about the country's economic problems, spoken out against corruption, and relayed citizen complaints to bureaucrats."[6] He has also supported privatization and diversification of the economy and called for the average Saudi to become less reliant on the government.

The Crown Prince and U.S.–Saudi Relations

Until recently, many in the Western press portrayed Crown Prince Abdullah as anti-American. When analysts described him, it was often with a sense of foreboding about what this powerful individual would mean to U.S.–Saudi relations. Many saw him "as a dark horse in the solidly pro-American ranks of senior Saudi princes."[7] But these fears have so far proved unfounded. As head the SANG, which has closer relations with the U.S. military than does the rest of the Saudi military, Crown Prince Abdullah understands the importance for Saudi Arabia of close cooperation with the United States. Although the crown prince places his country's strategic interests first, he also realizes that a solid relationship with the United States is essential to the kingdom's security and long-term goals. The crown prince emphasized this fact at a September 1998 meeting with U.S. oil executives, at which he termed U.S. oil companies "the bedrock" of U.S.–Saudi relations for half a century and said the "Saudi government wants them involved again in a new strategic energy partnership."[8] According to one meeting participant, the crown prince also said, "I have invited you because I realize that our countries' relations are deep and strategic, and in the long term, this fact will only increase in importance."[9]

This meeting with U.S. oil executives was a signal that the crown prince intended to consolidate the close relationship with the United States that his father, King Abdulaziz, and all subsequent Saudi kings had cultivated.[10] As Ronald Neumann, U.S. deputy assistant secretary of state for Near Eastern affairs, noted, "This move, and other carefully planned openings in the Saudi market, will definitely bring new dynamics to the [U.S.–Saudi] relationship, including more investment, closer economic ties, and technology sharing."[11]

Another example of the crown prince's intent to perpetu-ate the process of maintaining positive relations with the West includes working with Europe and the United States to resolve problems in the Arab world. For example, many international observers noted that the deportation from Libya of the sus-pects in the Lockerbie bombing could not have occurred without Saudi assistance.[12] At the urging of United Nations (UN) secretary general Kofi Annan, the crown prince used his influence with Mu'ammar Qaddafi to convince the Libyan leader to agree to a UN Security Council directive to surren-der the suspects. And although the United States blocked an immediate end to sanctions against Libya, it "welcomed the positive developments" coordinated by the Saudis.[13]

In Saudi relations with Iraq as well, the crown prince has shown the importance he places on Saudi Arabia's ties to the United States. Although the kingdom has its own historical, moral, and strategic reasons for opposing the Iraqi regime, a desire to maintain strong relations with the United States has always been an important element in its anti-Iraqi policy. For example, as the de facto head of state, the crown prince has not moved to oust U.S. troops from Saudi soil despite domes-tic opposition; rather, he has offered the use of the kingdom's airspace for attacks on Iraq[14] and has even called for the over-throw of Saddam Husayn.[15]

The crown prince does not support all aspects of U.S. policy toward Baghdad. In a December 1998 letter to Presi-dent Bill Clinton, the crown prince made clear his disagreement over the Desert Fox bombing campaign. "It created a situation of sympathy with the Iraqi regime," the crown prince argued, "that Saddam Hussein can use to achieve his objectives, and it has also fissured the international posi-tion" toward containing him.[16] The solicitous tone of the letter, as well as its frankness, however, revealed the strength of the friendship between the allies. For its part, the White House responded in kind. Although admitting that the United States did not agree with the conclusions of the letter, senior officials took "the views of the Saudi allies very seriously. Their advice carries tremendous weight in our decision-making on Iraq as well as all issues related to the Gulf region."[17] It is also important to note that in public statements the crown prince has chosen to emphasize the areas of agreement, rather than disagreement, with the United States regarding Iraq.

U.S. officials are optimistic about the future of U.S.–Saudi relations. "Not in a long time," one U.S. State Department official said, "have we seen so much excitement throughout the whole diplomatic community regarding the Saudis. There is a sense that real progress may be around the corner."[18] Much of the kingdom's leadership share the same spirit and have lined up behind the crown prince to support the continuation of that "special relationship" the United States and Saudi Arabia have shared for decades. "The senior princes have grown quite comfortable with their relationship to America, especially in the years after the [Persian] Gulf War," a former Saudi cabinet minister remarked.[19]

Regional Focus

Despite strong relations with the West, the crown prince is not oblivious to the kingdom's regional interests and responsibilities. As U.S. Deputy Assistant Secretary of State Neumann stated, "The crown prince is a deep, sincere Saudi nationalist."[20] Only by remembering this statement can the United States hope to understand the crown prince's domestic, regional, and international policies, and why he may risk relations with the United States when maintaining those good relations conflict with his other goals. "The crown prince is a strong and reliable friend," as former acting U.S. ambassador to Saudi Arabia David Welch observed. "But that does not mean that we will always see eye-to-eye, nor that Saudi Arabia is an American pupil."[21] As former U.S. ambassador to Saudi Arabia Walter Cutler bluntly stated, "We have to get it into our heads [that] the Saudis cannot really be bossed around."[22]

Issues on which the crown prince has diverged from U.S. policy include moving much more slowly to recognize Israel, and much more quickly on rapprochement with Iran, than the Clinton administration would have liked. In addition, no amount of wrangling by U.S. energy secretary Bill Richardson during his February 1999 visit to Saudi Arabia could convince the crown prince to open the kingdom's upstream oil sector to U.S. companies more quickly.

The policy positions of the crown prince that contradict U.S. policy are logical for strictly domestic reasons, but they also reflect his desire to strengthen Saudi Arabia's regional position. Domestic and foreign observers have been touting the crown prince's chances for becoming an important voice in the Arab world for many reasons: because other leaders in the region have been uninspiring, because he is a veteran leader at a moment of generational change throughout the region, and because the crown prince is widely respected and trusted and perceived as someone who has been able to maintain close ties to the United States without being viewed as subservient to Washington.

Delegation to Professionals

The crown prince's approach to decision making continues the Saudi evolution toward greater utilization of those who have technical expertise and professional experience, especially in the petroleum and banking sectors where Saudi expertise is relatively deep. As one former U.S. board member of Saudi Aramco noted with surprise, "From the start, Abdullah delegated authority to the experts, especially in the petroleum sector."[23]

This approach has characterized Crown Prince Abdullah's approach to governing since he first took the reigns of power in 1995. In fact, the crown prince's influence was demonstrated as early as the 1995 government cabinet shakeup, during which the crown prince appointed fifteen new members out of a total of twenty-nine cabinet positions. Western observers were surprised to discover that in general, the new faces were "noted technocrats with high educational qualifications."[24] In June 1999, after the customary four-year interval, the crown prince made a few more cabinet personnel changes, fulfilling his promise that his choices would once again "make use of Saudi expertise."[25]

The crown prince's management of Western oil company

proposals to invest in the kingdom's petroleum sector provides another example of his delegation principle. In the past, a small group of senior leaders would have studied proposals regarding the energy sector, or the oil minister would simply have made a decision by himself, as often happened during the tenures of former ministers Zaki Yamani and Hisham Nazer. By 1999, however, the situation changed. First, the crown prince, along with Foreign Minister Prince Saud Al Faisal, clearly indicated that these negotiations would take place between the oil company executives and the Saudi leadership directly, eliminating middlemen and the potential for kickbacks. The leadership would then assess the proposals only after a process of thorough analysis and commentary by many professionals.[26] In this manner, Saudi policymaking came closer to resembling that of any professional bureaucratic organization or industrialized nation.

The Next Generation

Crown Prince Abdullah has five sons who have entered public service. His eldest, Prince Khaled, was the former Saudi Arabian National Guard deputy commander for the Western Region and is currently a businessman in Riyadh. Prince Mitiab is the vice commander of the Saudi Arabian National Guard's military command and the president of the King Khaled Military Academy. The crown prince's third son, Prince Abdulaziz,

Crown Prince Abdullah's Sons in Official Positions	
Prince	**Position**
Khaled	Former SANG Deputy Commander for the Western Region; currently a businessman in Riyadh
Mitiab	SANG Vice Commander of the Military Command; President of King Khaled Military Academy
Abdulaziz	Adviser to the Crown Prince
Faisal	Private businessman
Turki	F-15 Pilot in the Royal Saudi Air Force

is currently an adviser in his father's cabinet, while Crown Prince Abdullah's fourth son, Faisal, is a private businessman. The crown prince's fifth son of adult age, Prince Turki, is an F-15 pilot in the Royal Saudi Air Force.

Prince Sultan, Second Vice Prime Minister; Minister of Defense and Civil Aviation; Inspector General

Prince Sultan, born in 1930, is the second full brother of King Fahd. He oversaw the creation of the Ministry of Agriculture in 1953 and moved in 1955 to establish the Ministry of Communication, planning the massive infrastructural development that the kingdom has experienced since that time. Currently, as minister of defense and civil aviation, he controls Saudi Arabian Airlines (formerly Saudia), the state-owned carrier.

As minister of defense, he oversees a quarter of the Saudi budget. Known as the father of the modern Saudi armed forces, he has overseen the development and modernization of the Saudi military since his appointment as minister of defense in 1962. Yet, Western observers admit that the Royal Saudi Land Forces (RSLF) and the Royal Saudi Navy (RSN) still have to make enormous progress before they are able to defend the vast borders of the kingdom adequately. In a 1999 Center for Strategic and International Studies report, Dr. Anthony Cordesman noted, "Saudi Arabia now has a massive backlog of [military] orders and serious training, conversion, [Operation and Maintenance], and sustainability problems."[27] Part of the explanation for these problems is that most of the military's attention has gone to upgrading the Royal Saudi Air Force (RSAF), which is, according to a former U.S. Air Force general who took part in the Gulf War that liberated Kuwait, the most modern and powerful of the major Arab air forces.[28] It should be noted, however, that whereas Prince Sultan has enlarged and modernized the Saudi armed forces, the kingdom's military has never posed a threat to its neighbors.

Regarding issues of national defense, Prince Sultan has been the principal actor in Saudi Arabia's "special relationship" with the United States. A spring visit in 1997 to Washington highlighted the importance of his position. He met with President Clinton and National Security Advisor Sandy Berger and participated in the establishment of a direct communications "hotline" between himself and Secretary of Defense William Cohen.[29]

Prince Sultan has also been active on foreign policy issues. For example, in May 1999 he became the first Saudi defense minister to visit Iran in three decades—a delicate undertaking, especially for a defense minister. He sidestepped the Iranian desire for military cooperation while emphasizing improvements in economic relations that he hoped could lead to better political relations between the two countries.[30] In the end, the prince managed to leave the Iranians pleased with the trip's outcome while garnering praise from Washington.

Concerning domestic economic and social issues, Prince Sultan has bluntly advocated opening state enterprises to the Saudi private sector, especially in the fields of power generation and aviation. Toward this goal, he supported the creation of the National Airline Corporation, a private airline that caters to the Saudi domestic market. Although he has also expressed interest in privatizing Saudi Arabian Airlines, he wants first to complete an internal reorganization of the company to make the privatization more successful.[31] Prince Sultan is also the founder and chairman of the Prince Sultan bin Abdulaziz Al Saud Foundation, which provides health, social, and educational services to Saudis. Programs include research grants, care for the elderly and handicapped, and plastic surgery for those with physical deformities.[32]

The Next Generation

Prince Sultan has six sons who have attained prominence in the Saudi government. The eldest, Prince Khaled, was the commander of the Saudi Air Defense Forces. He retired in 1991 after commanding all Arab forces in Operation Desert Storm. Prince Fahd is currently the governor of Tabuk Province, bordering Jordan. Prince Bandar is a former air force pilot who became the Saudi ambassador to the United States in 1983. Prince Faisal is the secretary general of the Prince Sultan Foundation for Charitable Works. Prince Turki is the deputy minister of information for foreign information affairs. Finally, Prince Salman is the assistant defense attaché in Washington, D.C.

Prince Sultan's Sons in Official Positions	
Prince	**Position**
Khaled	Former Commander of the Saudi Air Defense Forces
Fahd	Governor of Tabuk Province
Bandar	Saudi Ambassador to the United States
Faisal	Secretary General, Prince Sultan Foundation for Charitable Works
Turki	Deputy Minister of Information for Foreign Information Affairs
Salman	Assistant Defense Attaché in Washington

Notes

1. See Charts A, B, C, D, E, and F, in the appendices.

2. Faiza Saleh Ambah, "Crown Prince Popular with Saudis," Associated Press, August 4, 1999 (quoting Dr. Waheed Hashem, associate professor of political science at King Abdulaziz University in Jeddah).

3. Graham Fuller, former National Intelligence Council vice chairman at the Central Intelligence Agency, said in an interview with British Channel Four's television documentary series *Dispatches*, "I think it is difficult to say how long [the Saudi royal family] would survive . . . I could imagine the regime lasting for five to ten years even." See "The Saudi Tapes," *Dispatches*, British Channel Four Television, February 20, 1997 (quoting a November 1996 interview).

4. Crown Prince Abdullah's mother comes from the noble Al Rasheed family of the all-powerful Shammar tribe in north-central Saudi Arabia. The Shammar tribe is considered one of the kingdom's most prominent and powerful tribal confederations, and members of the Al Rasheed are its nominal leaders. Note also that the ages and birth years listed in this chapter may differ from Western sources; the sources used here are the Royal Saudi Archives in Riyadh.

5. David Hirst, "Fall of the House of Fahd," *Guardian* (London), August 11, 1999, p. 13.

6. Faiza Saleh Ambah, "Crown Prince Popular with Saudis," Associated Press, August 4, 1999.

7. John Rossant, "If Big Oil Pumps in Cash, Will It Solve Saudi Woes?" *Business Week*, November 2, 1998, p. 58.

8. Tom Doggett, "U.S. Oil Firms Hold Rare Meeting with Saudi Prince," Reuters, September 26, 1998.

9. Author interview with Nathaniel Kern, president of Foreign Reports, Inc., Washington, May 25, 1999.

10. Toby Odone, "Saudi Arabia: Al Apertura?" *Energy Compass* 9, no. 46, p. 6.

11. Author interview in Washington, May 5, 1999.

12. "Statement by HRH Prince Bandar Bin Sultan, Saudi ambassador to the United States, on the Lockerbie Case," *PRNewswire*, February 13, 1999, located online at http://www.prnewswire.com.

13. "Saudi Urges End to U.N. Sanctions on Libya," Reuters, July 10, 1999.

14. Steven Lee Meyers, "Attack on Iraq: The Overview; U.S. and Britain End Raids on Iraq, Calling Mission a Success," *New York Times*, December 20, 1998, Section 1, p. 1.

15. Saudi Press Agency, January 14, 1999.

16. Rowan Scarborough, "Saudi Letter Rejects Wisdom of Desert Fox Attack on Iraq," *Washington Times*, June 8, 1999, p. A3.

17. Ibid.

18. Author interview in Washington, January 16, 1999.

19. Author interview in London, March 3, 1999.

20. Author interview in Washington, May 12, 1999.

21. Author interview in Washington, May 26, 1999.

22. Author interview in Washington, May 27, 1999.

23. Author interview in London, March 3, 1999.

24. "Continuity of Policy Underlies Cabinet Shake-Up in Saudi Arabia," *Middle East Economic Survey* 38, no. 45 (August 7, 1995).

25. "Saudi Crown Prince Says Cabinet Reshuffle on Time," Reuters, June 1, 1999.

26. See "The Next Shock," *Economist*, March 6, 1999, p. 23. (describing abridged version of this process).

27. Anthony Cordesman, "Saudi Military Forces and the Gulf," *Middle East Studies Reports* (Washington: Center for Strategic and International Studies, February 4, 1999), p. 20.

28. Author interview in Washington, June 30, 1999.

29. Wahib Muhammad Ghurab and Muhammad Sadiq, "Prince Sultan Warns of 'Attrition Schemes' in Gulf," *al-Sharq al-Awsat*, March 2, 1997.

30. "Iranian, Saudi Defense Ministers Meet," Riyadh Saudi Arabian Television Network, May 2, 1999 (translated from Arabic by the Foreign Broadcast Information Service (FBIS)).

31. Muhammad Saman, "Private Airline Project Mooted," *al-Majallah* (London), April 5, 1999, pp. 45–48.

32. Information on the foundation is located online at http://www.sultancharity.org.

Chapter 3
Reinvigorating the Petroleum Sector:
The Role of Saudi Officials

The history of Saudi Arabia and oil go hand-in-hand. Ever since its discovery by the Arabian American Oil Company, a consortium of U.S. companies that had won a concession from King Abdulaziz in 1933, crude oil has defined the political and economic development of the kingdom. This assertion is not surprising considering the sheer volume of oil known to exist in Saudi Arabia: 260 billion barrels of proven reserves claimed in 1999, more than double the next nearest competitor, Iraq. At the end of the twentieth century, oil comprised 90 percent of Saudi Arabia's export earnings,[1] 75 percent of state revenues, and 35 percent of gross domestic product (GDP).[2]

This treasure is both a blessing and a curse to Saudi Arabia. On one hand, oil has provided the kingdom with the means to raise the standard of living and educational levels quickly and with much less sociopolitical disruption than most Arab states have experienced in the last fifty years. Oil has also made Saudi Arabia, with the largest economy in the Middle East, a regional powerhouse. On the other hand, few countries in the world are so reliant on a single source of income. In 1999, each $1 drop in the price of Saudi crude represented a $2.5-billion drop in Saudi annual income.[3]

The lens of the petroleum community therefore provides a clear picture of the changes occurring throughout the Saudi government. The kingdom's $45 billion–$50 billion annual energy industry is so significant and pervasive that it provides the best possible case study for demonstrating that Saudi Arabia has made enormous strides in developing a modern

17

and professional bureaucracy to administer the affairs of state.

Because oil policy has such an enormous effect on the political and economic health of the kingdom, it is not established by the whim of any individual but by the consensus of influential ruling family members after considerable debate and consultation with Saudi experts. Some argue that high prices should be maintained at all costs, mainly by cutting production; others maintain that achieving market share is the best strategy, through increasing production and lowering prices. The former strategy increases revenues in the near term, ensures that petroleum resources are not dissipated too quickly, and benefits other oil-producing nations. Drawbacks for the Saudis include the fact that higher prices inspire research into alternative fuels; bolster otherwise unprofitable production operations, thus adding to the number of competitors; and encourage exporters to cheat on export quotas established by the Organization of Petroleum Exporting Countries (OPEC)—that is, produce more to take advantage of the higher prices—thus lowering the market share of those producers who try to maintain the quota.

The second strategy—achieving market share—also has its advantages and drawbacks. Analysts note that if the Saudis simply ignored their associates in the Organization of Petroleum Exporting Countries (OPEC) and maximized their production capacity, oil prices would quickly drop to $5 per barrel but would nevertheless return a healthy 15 percent to the Saudis, whose production costs are so low.[4] Other analysts point out that after four to five years of lost income because of per barrel price decreases, low-cost Gulf Cooperation Council (GCC) producers, especially Saudi Arabia, would then see a substantial increase in revenue. The main problem is "find[ing] a way to bridge the four to five year revenue gap between the time they decide to shift" from a strategy of price maintenance to market share.[5]

Some argue that even if the Saudis discover a way to bridge the gap in revenue, a "flood-the-market" approach would infuriate and impoverish many OPEC members. They also point out that the economic disruptions to other oil-producing nations could severely upset regional stability and present a significant strategic threat to the kingdom. One cannot rule out this possibility because of the temptation for the Saudi leadership to eliminate the need to haggle with OPEC, drive high-cost producers from the market, and decimate alternative energy research with one swift action. This policy, however, would certainly require substantial foreign investment in the Saudi upstream petroleum industry, as maximum-capacity production would quickly consume the kingdom's spare production capacity.

Saudi oil minister Ali Al Naimi is the chief advocate for the strategy of maintaining a higher price despite having a lower market share, and the proponents of market-share-over-price strategy are led by deputy oil minister for petroleum policy Prince Abdulaziz bin Salman. Nevertheless, none of the leading players in the Saudi petroleum community are extremist proponents of either strategy. No one wants to flood the market; neither are there any price hawks. The differences that exist involve achieving the right "blend" of policy over time. Prince Abdulaziz, for example, was a principle architect of the March 1998 Riyadh pact establishing the higher price and lower market share idea as Saudi policy, even though he generally supports the opposite policy. Indeed, by mid-1999, the majority of petroleum officials had adopted a strategy to maintain prices in the $17–$20 barrel range for Brent benchmark crude oil.

Significantly, the petroleum community has debated and analyzed this question extensively over the last few years. Rather than simply adopting the position of one or two individuals, a long and thoughtful attempt to find the best approach ensued, indicating the maturation of the decision-making apparatus in the kingdom. The process, far from being dictatorial as many analysts assume, has actually been quite public and unrestrained.

Although a great number of individuals at many levels participate in what can be termed the Saudi petroleum community, this chapter will examine four main organizations: the Supreme Council for Petroleum and Minerals Affairs, the

Ministry of Petroleum and Mineral Resources, and the two
(soon-to-be-disbanded) ad hoc committees dedicated to evaluating foreign investment in the kingdom's energy sector. All
the members of these groups are Western-educated and, except for one, all are native Saudis. The analysis below will
demonstrate that another characteristic they hold in common is that they are all highly qualified to steer Saudi Arabia's
petroleum industry through the complex world of oil production, refining, marketing, and distribution.

Key Governmental Players in Saudi Petroleum Sector	
Crown Prince Abdullah	Vice Prime Minister and Commander of SANG
Prince Sultan	2nd Vice Prime Minister and Minister of Defense
Prince Saud Al Faisal	Minister of Foreign Affairs
Eng. Ali Al Naimi	Minister of Petroleum and Mineral Resources
Prince Abdulaziz bin Salman	Deputy Minister of Oil for Petroleum Affairs
Prince Faisal bin Turki	Senior Adviser, Oil Ministry
Ibrahim Khaberi	Deputy Minister of Oil for Mineral Resources Affairs
Abdelrahman Abdelkarim	Deputy Minister of Oil for Corporations
Sulaiman Al Herbish	Governor to OPEC (rank of Deputy Minister)
Dr. Ousama Tarabulsi	Director of Auditing Affairs (rank of Deputy Minister)
Dr. Mohammad Al Saban	Senior Economic Adviser in Oil Ministry
Dr. Majed Al Muneef	Senior Economic Adviser in Oil Ministry
Dr. Saud Al Ammari	General Counsel to the Oil Minister

The Supreme Council for Petroleum and Minerals Affairs

The Supreme Council for Petroleum and Minerals Affairs was reinvigorated on January 4, 2000, by the first Royal Decree of the new millennium. The new council will have a greater legal authority than the former Supreme Petroleum Council and will have under its jurisdiction all governmental organs pertaining to the petroleum community—the Ministry of Petroleum and Mineral Resources, the Petroleum Ministerial Committee, the Petroleum Preparatory Committee, and Saudi Aramco—as well as the country's industrial base, including Saudi Arabian Basic Industries Corporation (SABIC) and the Saudi Electric Company (SEC).

The council will have the wide-ranging mission of overseeing and determining oil and gas production levels; it will control and direct all aspects of the oil, gas, and minerals industry; and it will formulate all policies. It will directly oversee the Saudi Arabian Oil Company—Saudi Aramco—and approve its board of directors. The council will also determine domestic oil pricing and deal with other international and national marketing questions. Moreover, the new Supreme Council will be the sole governmental body to determine and grant any future joint venture agreements with international oil companies (IOCs). The two ad hoc committees mentioned above functioned in an advisory capacity only, as the Supreme Council will have the final say on what ideas and projects will be proposed to the IOCs. Once the final report and all relevant background notes have been written and handed to the Supreme Council, the two ad hoc committees will be completely disbanded.

The establishment of this new council demonstrates the keenness of the crown prince to incorporate in the decision-making process all the major government experts. As a senior Shura (Saudi consultative council) member noted, Crown Prince Abdullah feels that "delegating authority to the appropriate group of people with the necessary know-how is the only way to successfully revamp and reform the national economy. The new Supreme Council perfectly demonstrates his strong will and desire to do so."[6]

In late February 2000, the Saudi government also created a Specialized Committee on Foreign Investments to work as a subgroup of the Supreme Council. This committee will handle the reentry of IOCs into the kingdom's petroleum sector.

Members of the Supreme Council

The Supreme Council is to be chaired by the king; the crown prince will be the vice chairman, and Prince Sultan, the minister of defense and civil aviation, will be the second vice chairman. The other nine members are as follows:

Prince Saud Al Faisal, Minister of Foreign Affairs. Prince Saud Al Faisal, son of the late King Faisal, was born in 1942 and graduated from Princeton University with an economics degree in 1964. He served as the deputy governor of Petromin in 1970 before becoming deputy minister of oil for petroleum policy in 1971. He remained in that position until 1975, when he was named foreign minister. As a result of his experience in the field, he understands Saudi Arabia's foreign policy prerogatives and the international oil industry extremely well and has emerged as the crown prince's chief adviser on foreign affairs and energy policy.

According to a former Saudi cabinet member, Prince Saud "has a reputation as a skillful negotiator and diplomat." For example, he was instrumental in orchestrating the Taif accords of 1989,[7] which ended fifteen years of civil war in Lebanon by granting Sunni and Shi'i Muslims more political power, reforming a system which had kept Christians dominant long after demographic change had reduced them to a minority. By 1996, according to one special report, "Saudi Foreign Minister Prince Saud Al Faisal and the other sons of the late King Faisal [had] gained an impressive reputation for their hard work and honesty."[8] As a result, he has the ear of the crown prince and the other senior princes.

In fact, Prince Saud is generally acknowledged as one of the leading third-generation princes. A recent British ambassador to Saudi Arabia commented, "Due to his strong aptitude and perfectionist mentality, the Saudi foreign minister has clearly emerged as the most powerful and influential third-generation prince in the Royal House of Saud."[9] Likewise, a recent French ambassador added, "There are no doubts for [the French government] that [Prince] Saud has become over the last three years the man to go to when vital strategic issues are to be discussed."[10]

Prince Saud seems to acknowledge the need for foreign investment and closer economic cooperation between the kingdom and the world's major industrialized powers. He favors opening Saudi Arabia's energy sector but apparently believes that this process must occur in a cautious manner, always keeping the kingdom's interests firmly in mind. Specifically, he reportedly favors opening the country's gas and electrical sector and increasing the stake of U.S. companies in the downstream petroleum sector. Yet, he was one of the first to see that opening the kingdom's upstream sector to foreign investment did not make economic sense in the short term.

In many ways, Prince Saud is one of the most effective bridges between the kingdom and the West. He defends Saudi priorities but understands the U.S. mindset better than many other senior government officials. Former U.S. ambassador to Saudi Arabia Walter Cutler has noted that Prince Saud often acts as a gatekeeper to the Saudi leadership because "he is quite adept at translating American thoughts and ideas into a Saudi cultural context, and vice versa."[11]

Prince Saud will also serve as the chairman of the Supreme Council's Specialized Committee on Foreign Investments.

Eng. Ali Al Naimi, Minister of Petroleum and Mineral Resources. Ali Al Naimi was born in 1935 and began working at Aramco at the age of 12, rising through the ranks to become the first Saudi chief executive officer (CEO) of that company. Al Naimi is Western-educated, with a masters degree in geology from Stanford, and in some ways can be considered a Western-leaning manager. His education and experience under U.S. managers at Aramco influence his management techniques, and, when devising policy, he adopts the perspective of a U.S. executive. In other words, he bases decisions on technical and business factors rather than on politics, delegates authority to subordinates rather than closely holding information and power, and evaluates individuals primarily on their business skills rather than on their family backgrounds.

Al Naimi is a career industry man, not a political appointee—a fact the Western press has recognized, praising him as an "internationally respected oil technocrat"[12] and a "shrewd technocrat with vast oil industry experience."[13] Those who work closely with Al Naimi remark that he is a "tough negotiator who carefully measures his words . . . [and he] is always in full control of his feelings."[14]

Al Naimi is often blamed for recommending the ill-fated decision to increase output at 1997's Jakarta OPEC summit. Soon after the decision, Asian demand dropped drastically and oil prices began their steep decline. Naturally, as oil minister, he was blamed for the sharp decrease in the kingdom's oil revenues. He was not, however, alone among the international petroleum and financial industries in failing to predict the coming Asian crisis. On a positive note, however, a senior Saudi petroleum policymaker noted that Al Naimi "was the first, along with his assistants—most notably Princes Abdulaziz bin Salman and Faisal bin Turki—to bring about remedies and policy alternatives to live with the grave economic crisis caused by the slump in oil prices."[15]

In addition to the criticism he received over the price drop, he has often had chilly relations with former colleagues at Saudi Aramco. Western analysts familiar with the internal workings of Saudi Aramco argue that this tough management style, and the fact that he still maintains influence at Saudi Aramco after having formally moved to the position of oil minister, has created this friction among senior executives at the state oil company. One former American Saudi Aramco executive admitted, "Not everyone appreciates Naimi's weekly visits with [CEO] Abdallah Jumah."[16] For instance, Finance and Relations Executive Vice President Nabil Al Bassam retired eighteen months early in apparent protest of Al Naimi's "micromanagement."[17] But senior Saudi oil policymakers argue that, on the contrary, Al Naimi's retention of influence at the national oil company is a wise and expedient measure because it keeps up communication between the government and the country's most important enterprise.[18] A senior Aramco executive summarized the situation by saying, "The minister is a father-like figure to Saudi Aramco and its management. Hence, he has the legitimacy and influence to impose any decision on the oil company as he deems fit, and is still our most formidable protector."[19]

As members of the "old guard," Al Naimi and other senior executives at Aramco are wary of opening Saudi Arabia's upstream oil fields to foreign investment. "Mr. Al Naimi doesn't want to let the IOCs do what he thinks Aramco could do better," according to a former oil official who worked with Saudi Aramco.[20] This is a natural reaction of a man who is proud of the success that Saudis themselves have had in leading the Saudi Arabian Oil Company to its present stature, and fearful of the impact of international companies returning to the kingdom. Al Naimi's hesitation is also based on the economic reality that, as long as there is spare production capacity, increased investment in exploration and production is hard to justify. Changes in this stance, however, were seen as 1999 progressed, and given that by 2020 the kingdom could be forced to add between 3.5 million and 5 million barrels per day to production capacity, private foreign investment may be inevitable. "We will expand our oil servicing industry and link it to the private sector," the former oil official said, "and this includes exploration and other stages of oil production and related industries."[21]

Al Naimi's expertise is not limited to petroleum, though. More than thirty-five years ago, he wrote "Water Resources Development in North Eastern Saudi Arabia," which he presented in March 1965 at the Arab League Energy Conference in Cairo; a former U.S. Aramco senior executive to this day calls the paper "the leading study on water resources development in the Eastern Province."[22]

Al Naimi will serve as a member of the Specialized Committee as well as on the Supreme Council.

Dr. Ibrahim Al Assaf, Minister of Finance and National Economy. With a doctorate in economics from Colorado State University, Dr. Ibrahim Al Assaf worked as an adviser to the

Saudi Development Fund; professor at the Saudi Land Forces College; Saudi Arabia's executive director at the World Bank; deputy governor of the Saudi Arabian Monetary Agency (SAMA) from 1995 to 1996; and, finally, minister of state in the Saudi cabinet. [23] Upon his appointment to the cabinet in 1996, he commented that "the economic atmosphere is appropriate for the private sector to play a higher role in the national economy, given the opportunities presented by a strong and modern infrastructure and liberalization."[24] In addition to supporting economic liberalization, he has recognized the need for foreign investment in Saudi Arabia. According to former Deputy Assistant Secretary of the Treasury for Arabian Affairs Charles Schotta, Al Assaf supports foreign investments in the kingdom but "is also a proponent of ensuring that the process is conducted carefully and only after much deliberation."[25] This statement may sound like a diplomatic way of saying that foreign investment is an impossibility, but, in Saudi Arabia it is the stance of one who believes in a gradual liberalization of the economy but does not want the kingdom to suffer the same disaster that rash privatization brought to Russia.

Al Assaf has also spoken specifically of the benefits of ultimately opening both the kingdom's upstream and downstream petroleum sectors to foreign participation. This attitude, combined with his long experience and international perspective, has made him, according to a former senior Saudi Treasury Ministry official, a prototype of the "new generation of technocrats that are well-versed in development economics."[26] He is also a member of the Aramco board of directors, and he was recently named a member of the new Specialized Committee on Foreign Investments.

Dr. Hashim Yamani, Minister of Industry and Electricity. Yamani was appointed to his current position in 1995, and he was recently named a member of the Specialized Committee. He entered government from the academic sector, where he was a professor of physics at the King Fahd University of Petroleum and Mineral Resources. Yamani graduated from Harvard University's Graduate School of Arts and Sciences with a doctorate in physics. As the minister of industry and electricity, he oversees all the electric plants in the kingdom, as well the world's petrochemical giant SABIC, of which he is also the chairman. Hence, it is only natural for him to be a member of this council, especially when the kingdom will need over $120 billion in investments for the electric sector in the next two decades.

Yamani greatly favors an increased pace of foreign investments, especially in the electric, petrochemical, and heavy industries sectors, but he insists that the domestic administrative infrastructure be established before negotiations with the IOCs begin. He can be credited for successfully unifying all the regional Saudi consolidated electric companies (SCECOs) into the Saudi Electric Company, with an initial market capitalization of $8.9 billion.

Khaled Al Ghosaibi, Minister of Planning and Acting Minister of PTT. Al Ghosaibi was appointed to the Saudi cabinet in 1999 and was given the post, telephone, and telegraph (PTT) portfolio that December. Prior to his ministerial nomination, Al Ghosaibi was a prominent member of the Consultative Shura Council. Although he does not have much experience in the executive branch of the government, Al Ghosaibi had a long career in the private sector. Thus, he will be able to pay special attention to the interests and desires of the business establishment as the privatization process proceeds. He is also a supporter of market economics. Al Ghosaibi is the fifth member of the Specialized Committee on Foreign Investments.

Dr. Mutlib Al Nafisah, Minister of State and Secretary General of the Supreme Council. Al Nafisah became a member of the Saudi cabinet in 1995. Prior to this promotion, he was the chief of the Committee of Legal Experts in the Council of Ministers. He received a doctoral degree from Harvard Law School in comparative legal systems. Although a minister of state without portfolio, he is in charge of all international legal issues pertaining to the Saudi Arabian government, such as the ongoing discussions to demarcate the Saudi–Yemeni borders. As secretary general of the Supreme Council, Al Nafisah will set its agenda and play a central role in all the council's future decisions. Like his colleagues, he is known to favor more foreign direct investments in key sectors of the Saudi economy.

Dr. Saleh Al Adhel, President, King Abdulaziz City for Science and Technology. Dr. Saleh Al Adhel has been spearheading the kingdom's programs in scientific technological development, including alternative advanced technologies for the petroleum sector. He received a scholarship from the King Abdulaziz Society for Gifted Persons to study at Stanford University, where he received his doctorate in mathematics with a concentration in advanced trigonometric and algebraic equations. His expertise on the council will be in recommending what policies to adopt for the long-term technological enhancement of Saudi Arabia's oil, gas, and power sectors. According to a founding member of the King Abdulziz Society for Gifted Persons, "Dr. Saleh's special intellectual gifts, especially in numerical matters, will permit him to give a long-term, clear picture to the council of what will be the end results to any policy that it will adopt and implement in the near future."[27]

Abdulaziz Al Rashid, Former Deputy Minister of Finance and National Economy for Auditing Affairs. Mr. Abdulaziz Al Rashid is an accountant by profession. Although currently in private business, he had a long career in the Saudi Ministry of Finance and National Economy, where, according to a Saudi private banker, "He did a good job in keeping perfect records of revenues and expenditures."[28] Al Rashid's role in the council will be to keep track of all financial transactions undertaken by the various organs of the government that constitute the Saudi petroleum community. His focus will be on Saudi Aramco, as the council will be exploring different venues to trim the national oil company's huge expenditures.

Abdallah Jumah, President and CEO of Saudi Aramco. Abdallah Jumah serves as Aramco's CEO and sits on its board of directors. He joined the company in 1968, rose to the post of vice president for governmental affairs, and finally replaced Ali Al Naimi when Al Naimi became oil minister in November 1995. Jumah remains close to Al Naimi, even though the latter is no longer with Aramco. Jumah studied political science and graduated from the American University of Beirut, after which he completed the management development program at Harvard University in 1976.[29] He is known in the company for his characteristically U.S. management style.

Jumah's management skills manifested themselves during his tenure as vice president for governmental affairs. His responsibilities included working closely on joint projects with various ministries and government agencies, such as the Saudi Consolidated Electric Companies (SCECOs). In this capacity, according to a former Aramco executive, "he was able to develop good relations with people he interacted with, a skill which often helped him to solve problems before they threatened to delay a project or escalate into major crises."[30] In addition, as a current Aramco official noted, his work managing relations between Aramco and SABIC earned him a reputation as a consensus maker and team builder.

The Ministry of Petroleum and Mineral Resources

The Ministry of Petroleum and Mineral Resources is one of twenty-two Saudi government ministries and the center of the country's energy community. Currently headed by Oil Minister Al Naimi, it is responsible for implementing all of the kingdom's petroleum projects and policies. In addition to executing policy, almost all initiatives and ideas regarding petroleum policy originate in this institution. For these reasons, it is one of the most powerful ministries in the kingdom, and arguably, the most important.

The oil minister works with four primary deputy ministers, whose jurisdictions are petroleum policy, natural minerals, joint ventures, and administration. The deputy for petroleum policy is by far the most important and prestigious position of the four deputy minister slots, and it is generally regarded as the number two position within both the ministry and the petroleum community as a whole. As of late 1999, career energy technocrats who have spent their professional lives in the Oil Ministry filled these posts. After analyzing economic and geologic data, the Ministry's technocrats prepare recommendations that are apparently quite important in shaping policy. Although the oil minister decides many matters himself, he brings the most important policy issues to the Council of Ministers or directly to the king.

The kingdom's representative to OPEC, a position equivalent to deputy minister, reports directly to the oil minister, as does the Saudi representative to the OPEC Economic Commission. Finally, the state oil company, Saudi Aramco, also comes under the jurisdiction of this powerful ministry. As chairman of Aramco's board of directors, the oil minister directly supervises Aramco's president and CEO.

The Ad Hoc Committees

Operating alongside the Oil Ministry were two ad hoc committees created in late 1998 to receive, study, and evaluate foreign investment proposals in the kingdom's energy sector. These committees are known as the Petroleum Ministerial Committee and the Petroleum Preparatory Committee. Both committees are composed of some of the kingdom's most experienced technocrats. The committees played a major part in the initial stages of the process to reinvite the IOCs back into the kingdom, but their legal existence was superseded with the creation of the Supreme Petroleum Council. Many of the individuals on the ad hoc committees were nominated to the new council, but those who were not will continue to play an informal advisory role until the final report and all relevant background notes are handed to Crown Prince Abdullah for final deliberation.

The Petroleum Ministerial Committee

The Petroleum Ministerial Committee was formed in December 1998, two months after the crown prince's request for investment plans from U.S. oil companies. This invitation was the result of both long- and short-term trends. The long-term factors included the desire for closer economic relations with the United States; more capital to meet the country's development needs; and the advantages that foreign investment had brought to other countries in the Gulf, such as technical and marketing expertise. The immediate cause was the drastic fall in oil prices during 1998 that made it even more difficult for the kingdom to finance much needed domestic projects, especially those related to power generation and

Members of the Petroleum Ministerial Committee	
Name	**Other Positions Held**
Prince Saud Al Faisal (chairman)	Minister of Foreign Affairs
Ali Al Naimi	Minister of Petroleum and Mineral Resources
Dr. Ibrahim Al Assaf	Minister of Finance and National Economy
Dr. Mohammed Al Sheikh	Minister of State
Dr. Musaed Al Ayban	Minister of State
Dr. Abdulrahman Al Suhabani*	Economic Adviser at the Royal Court

* Dr. Al Suhabani is secretary general of the committee, but this position is purely administrative; he has no input or say in the final recommendations that the members of the committee present to the crown prince.

water desalination. Thus, at a September 1998 meeting in Washington, the crown prince asked U.S. oil executives to submit proposals for investment in Saudi Arabia's oil sector, and the committees were convened soon thereafter. From the beginning, the regent had strong backing from the senior princes and other leaders for this action.

The Ministerial Committee brings together the most qualified individuals from various elements in the state to oversee foreign involvement in the kingdom's energy sector. Its composition indicates that the committee, for its duration, will have real influence. The foreign minister is the chairman, and the ministers of petroleum and finance sit on the committee, along with two other ministers of state. This array of individuals from the major power centers in the government indicates the importance the senior Saudi leadership places on the question of opening up the kingdom's energy sector to foreign investment. Some of the members represent the emerging, young technocratic elite, while others represent the older, more experienced generation.

All committee members are free-market economists who understand the importance of foreign investment and were educated in the United States. Their main purpose on the committees is not necessarily to advocate certain policies but to maintain communication between the crown prince and his main advisers, along with the groups they represent. Thus, they play a role in helping to achieve consensus within the kingdom—a vital prerequisite for the successful implementation of any policy of national importance.

The committee chairman, Prince Saud Al Faisal, and members Ali Al Naimi and Ibrahim Al Assaf, were mentioned earlier in this chapter in the discussion of the Supreme Council for Petroleum and Minerals Affairs; the other two "voting" members of the Ministerial Committee are as follows:

Dr. Mohammed Al Sheikh, Minister of State. Dr. Mohammed Al Sheikh, former minister of urban and social affairs, has a doctorate in civil engineering from the University of California–Berkeley. According to a senior member of the Consultative Council, Al Sheikh's successful reorganization and revitalization of this ministry in the early 1990s elevated his profile in economic and social affairs, and he has become a leading voice on economic reform.[31] Regarding privatization, he has indicated his skepticism of completely divesting the state's assets, apparently because he believes that significant social reforms must proceed privatization in order to ensure a successful transition from public- to private-sector control.[32] According to one Saudi ambassador in Western Europe, Al Sheikh has also argued that—based on the current, international oil-market situation and the kingdom's spare production capacity—Saudi Arabia does not need foreign involvement in its energy sector in the short term.

It is also important to note that the Al Sheikh family, as descendants of Sheikh Muhammad ibn 'Abdul Wahhab (1703–92), is the preeminent spiritual authority in Saudi Arabia.

Dr. Musaed Al Ayban, Minister of State. The youngest member of the cabinet since his appointment in 1996, Dr. Musaed Al Ayban is one of eight ministers of state who sit on the Council of Ministers and engage in special tasks that the king and the crown prince deem important; this included sitting on the Petroleum Ministerial Committee in 1998 and 1999. Possessing a doctorate from Harvard Law School and a strong background in legal issues, he brings a legal perspective to the work of the committee, working out the details of proposals to ensure that they are mutually beneficial to the IOCs and to the kingdom. It is important to note that in late 1999 he came out against the idea of foreign involvement in the upstream sector.

Al Ayban is also a member of the Aramco Board of Directors. Prior to being named a minister of state, he was a legal consultant at the Ministry of Petroleum and Mineral Resources, while concurrently an assistant professor at King Saud University.

Petroleum Preparatory Committee

Immediately after the convocation of the Ministerial Committee, the Petroleum Preparatory Committee was created to receive and initially evaluate proposals from the IOCs. Similar to the Ministerial Committee, the Preparatory Committee consists of representatives from the major governmental institutions—finance, oil, foreign affairs, and the royal court.

Prince Abdulaziz bin Salman, Deputy Minister of Oil for Petroleum Affairs. The son of the charismatic governor of Riyadh, Prince Abdulaziz bin Salman exemplifies the third generation of royal princes in high government positions. Before assuming his post as deputy minister, he had been an adviser in the Ministry of Petroleum and Mineral Resources for ten years. By the end of the 1990s, he had gained a solid background in the oil industry and was the primary architect of the 1998 talks, among OPEC members Saudi Arabia and Venezuela and non-OPEC member Mexico, which laid the groundwork for substantial production cuts.[33] Prince Abdulaziz completed his undergraduate and graduate studies at the King Fahd University of Petroleum and Mineral Resources. After completing his studies, he worked at the university's research institute and launched its Center for

Members of the Petroleum Preparatory Committee	
Name	**Other Positions Held**
Prince Abdulaziz bin Salman (chair)	Deputy Minister of Oil for Petroleum Policy
Dr. Youssef Al Sadoun	Deputy Minister of Foreign Affairs for Economic and Cultural Affairs
Saleh Al Naim	Director General, Saudi Industrial Development Fund
Dr. Abdel-Rahman Al Suhabani*	Economic Adviser, Royal Court

* Dr. Al Suhabani is secretary general of the committee, but as with the Ministerial Committee, this position is purely administrative; he has no input or say in the final recommendations that the members of the committee present to the crown prince.

Energy Economics Research, and he also took some courses in U.S. foreign policy at the School of Advanced International Studies (SAIS) at the Johns Hopkins University.

In addition to working on means of cooperation in the petroleum sector while traveling in the West, Prince Abdulaziz has been active in promoting Saudi–Asian relations. For example, he is currently spearheading the Saudi initiative toward the East Asian economic giants—Japan, China, and South Korea. During the latter years of the 1990s, he spent much of his time meeting with business leaders and government officials in these countries, seeking ways to expand Saudi Arabia's business relations with them. He is also the main Saudi negotiator with the Japanese on the possible renewal or extension of the Divided Zone Concession to the predominantly Japanese-owned Arabian Oil Company.

Prince Abdulaziz is watched closely by Western analysts. As a former Saudi cabinet minister said, he is "a strong candidate for having a larger say in formulating Saudi oil policy in the future, since his prestige and power will increase as he gains more experience." [34] During the March 1998 meeting he arranged in Riyadh, Venezuela and Mexico agreed, along with Saudi Arabia, to cut production. This agreement marked the beginning of a turnaround for low oil prices that ultimately gave rise to the overall OPEC agreement in March 1999 to decrease production and thus increase oil prices. As a possible future oil minister, he can be expected to perpetuate the liberalizing economic policies of his predecessors or even to move them further toward free market ideas. For example, he is a strong supporter of the notion that Saudi Arabia must be open-minded toward foreign investment in the upstream sector of the petroleum industry, focusing on the commercial aspects and long-term viability of proposals. This position apparently is based on his strong belief in the importance of achieving market share—a policy requiring the kingdom to expand its spare production capacity in the mid-term and probably also requiring foreign capital to implement successfully.

In the late 1980s, Prince Abdulaziz opened talks with the Paris-based International Energy Agency (IEA), an organization established in response to OPEC to orchestrate the energy policies of consumer nations. Through these talks, he eventually paved the way for communication between OPEC and the consumer nations. Thus, he was in essence the first to start the producer–consumer dialogue that culminated in cooperation during the Gulf crisis. During this time, he maintained secret contacts with the IEA, to help coordinate the policies of OPEC and the main consumers. This cooperation was one of the reasons that OPEC quickly increased its production to fill the supply vacuum left by Kuwait and Iraq during the Gulf War. Prince Abdulaziz was also the first to set up the ongoing dialogue between the U.S. Department of Energy and the Saudi Oil Ministry. As Ambassador William Ramsay, deputy executive director of the IEA, has said, "Prince Abdulaziz has played a crucial role in opening lines of communication between the main producing and main consuming oil countries, especially between the kingdom and the United States." [35] This cooperation was best seen during the Gulf War. Ambassador Ramsay, who was previously the deputy assistant secretary of state in the Bureau of Economic and Business Affairs, added, "During my last tenure at the State Department, which was at the time of the Gulf War, Prince Abdulaziz was a critical contact for us [the consuming nations]. He not only took part in coordinating Saudi Arabia's immediate production increase, but he also took personal responsibility for assuring that all Allied military forces operating out of Saudi Arabia never ran out of fuel supplies. At that time, this was not an easy task at all." [36]

Dr. Youssef Al Sadoun, Deputy Minister of Foreign Affairs for Economic and Cultural Affairs. Another Western-trained professional in public policy, Dr. Youssef Al Sadoun represents the foreign policy establishment on this committee. With a doctorate in public administration from the University of Pittsburgh, he is known to be a free-market economist and firm supporter of opening the kingdom to more Western investment and cooperation. He is one of the foreign minister's primary advisers on international economic issues and has become one of the principal Saudi negotiators in discussions with the World Trade Organization (WTO) for Saudi accession.

Saleh Al Naim, Director General of the Saudi Industrial Development Fund (SIDF). Saleh Al Naim served an advisory role in the committee, but he was not appointed to the Supreme Petroleum Council.

Other Key Officials in the Petroleum Community

Prince Faisal bin Turki, Senior Adviser in the Oil Ministry. The son of Prince Turki bin Abdulaziz, former vice minister of defense and civil aviation, Prince Faisal has been in the Oil Ministry for approximately a decade. He has a bachelor's degree from the King Fahd University of Petroleum and Mineral Resources and shoulders several responsibilities in his current position. Primarily, he is the ministry's pointman in dealing with the domestic market and the local industrial energy-consuming giants such as Saudi Arabian Basic Industries Corporation (SABIC), the Saudi Consolidated Electric Companies (SCECOs), and the Saline Water Conversion Corporation (SWCC). He enjoys a close working relationship with his cousin, Prince Abdulaziz bin Salman. According to an American executive at Aramco, "Prince Faisal is emerging as a perfect partner to his cousin, the deputy minister, on the domestic scene. He has clearly become the number three man at the ministry."[37]

Sulaiman Al Herbish, Saudi Governor to OPEC. Sulaiman Al Herbish joined the Oil Ministry in 1962 after graduating from Cairo University.[38] Al Herbish has been the Saudi OPEC governor since 1990 and is currently the chairman of both the Arab Drilling Company and Saudi Arabian Texaco.[39] His experience with OPEC and deep involvement in the production-cut negotiations over the past two years made him a strong choice for OPEC secretary general, a nomination he received in June 1999.[40]

Notes

1. Michael Georgy, "Saudi Desert Offers Oil Investment Guide," Reuters, February 22, 1999.

2. The Energy Intelligence Group (EIG) placed the estimate of the share in GDP at 40 percent; see EIG, "Saudi Aramco," part of the series *The World's Key National Oil Companies,* January 1999, p. 8.

3. Georgy, "Saudi Desert Offers Oil Investment Guide," February 22, 1999.

4. According to one economist, this would be a "mid-term rate of return"; see "The Next Shock," *Economist,* March 6, 1999, p. 22.

5. Dr. H. Franssen, "Oil Price Developments and Investment Opportunities," presentation at the Ninth Annual Conference of the Centre for Global Energy Studies, London, April 1999.

6. Author interview in Riyadh, January 7, 2000.

7. Taif is the Saudi summer capital, located in the Western province of Hejaz.

8. Dr. Andrew Rathmell and Dr. Mustafa Alani, "Saudi Arabia: The Threat from Within," Special Report no. 12, *Jane's Intelligence Review,* November 1996.

9. Author interview in Geneva, February 22, 2000.

10. Author interview in Evian, France, February 19, 2000.

11. Author interview in Washington, May 27, 1999.

12. "Continuity of Policy Underlies Cabinet Shake-Up in Saudi Arabia," *Middle East Economic Survey* 38, no. 45 (August 7, 1995), located online at http://www.mees.com/archive/volumes/volume38/v38n45/38n45b01.htm.

13. Michael Georgy, "Naimi Gets Vote for Steady Saudi Oil Policy," Reuters, June 16, 1999.

14. Ibid.

15. Author interview in Riyadh, November 5, 1999.

16. Author interview in Washington, July 2, 1999.

17. "Saudi Aramco Restructures after VP Departure," *Energy Compass*, November 27, 1998, p. 9.

18. Author interview in Riyadh, October 12, 1999.

19. Author interview in Dammam, Saudi Arabia, November 2, 1999.

20. *Petroleum Intelligence Weekly*, May 3, 1999, p. 1.

21. Ibid.

22. Author interview in London, April 12, 1999.

23. "King Fahd Appoints 'Assaf Finance and Economy Minister," *Middle East Economic Survey* 39, no. 19 (February 5, 1996).

24. Ibid.

25. Author interview in Washington, June 11, 1999.

26. "King Fahd Appoints 'Assaf Finance and Economy Minister," *Middle East Economic Survey (MEES)*, February 5, 1996, p. 7.

27. Author interview in Riyadh, February 4, 2000

28. Author interview in Manama, Bahrain, January 28, 2000

29. See the Saudi Aramco Website, online at http://www.aramco.com/aboutus/corpmgmt.html.

30. Author interview in London, April 12, 1999.

31. Author interview in Geneva, August 28, 1999.

32. Ibid.

33. Nathaniel Kern, interview with author in Washington, May 5, 1999.

34. Author interview in Los Angeles, January 1999.

35. Author interview in Paris, September 15, 1999.

36. Ibid.

37. Author interview in Riyadh, October 25, 1999.

38. "Iran to Submit Candidate to Rival Saudi for OPEC's Top Job," Reuters, June 27, 1999.

39. "Saudi Arabia Nominates Oil Executive for OPEC Secretary General," Bloomberg, June 21, 1999.

40. Ibid. Although the decision was to be made at the September 1999 OPEC conference, the Iranians and Iraqis did not support Al Herbish, so the vote was shelved until the subsequent meeting in March 2000; the election of OPEC secretaries general must be unanimous.

Chapter 4

The Petroleum Sector
and Saudi Aramco

The Saudi Arabian Oil Company (Aramco) is one of the most powerful oil companies in the world. It currently manages the largest-proven petroleum reserves on Earth, nearly twelve times that of the United States, and is the world's largest oil producer and exporter. It is also the world's leading producer of liquefied petroleum gas (LPG) and the ninth largest producer of natural gas. With 56,000 employees (80 percent of all employees and 95 percent of senior management are Saudis)[1] and operations that span Asia, Europe, and the United States, the company is a global powerhouse. It runs eleven medical clinics, recreational facilities for its employees, and "one of the world's largest privately funded industrial training programs."[2] By controlling the most important resource of the kingdom, the firm is said by some observers to be "the primary instrument of the state in pursuing its foreign policy and economic objectives."[3] Specifically, it is a critical tool in shaping oil prices (keeping them neither too high nor too low), exercising power within OPEC, and achieving Saudi foreign policy objectives.

In addition to its enormous operations and assets, the company has unequaled spare petroleum production capacity. This capacity has allowed the Saudis to act as the world's "swing producer" for more than twenty-five years, helping to stabilize world prices through several regional wars and economic upheavals. For example, during the Persian Gulf War of 1990, when Iraq's invasion of neighboring Kuwait removed 4 million barrels per day (bpd) from the world's market,[4] the kingdom soon increased its own production to replace much of the difference. After the March 1999 production cuts, when Saudi production dropped to 7.5 million bpd, the state possessed around 3 million barrels of spare production capacity—enough to send terror into even the most determined OPEC "quotabuster." (See the table below and Chart 1 in the appendix section.)

OPEC Spare Capacity, March 1999				
	Total Capacity (000s)	Target Output (000s)	Capacity Utilization (percent)	Spare Capacity (000s)
Saudi Arabia	10,500	7,438	69%	3,362
Iran	3,700	3,359	91%	341
Venezuela	3,600	2,720	76%	880
UAE	2,650	2,000	75%	650
Kuwait	2,650	1,836	69%	814
Nigeria	2,320	1,885	81%	435
Indonesia	1,420	1,187	84%	233
Libya	1,560	1,227	79%	333
Algeria	930	731	79%	199
Qatar	740	593	80%	147
TOTAL	**30,070**	**22,976**	**76%**	**7,394**

Notes: (1) By June, 1999, OPEC compliance reached 90 percent. (2) Iraq is excluded from this table because it is likely to continue producing at capacity for the near term. (3) There is almost no unused capacity outside of OPEC, except for Norway, Mexico, and Oman.
Sources: Reuters, BT Alex Brown, Bloomberg, and the author's personal calculations.

The company is governed by a board of directors and chaired by the oil minister. Senior public servants with backgrounds in economics, political science, and petroleum issues comprise the board. The directors come from many sectors of the kingdom: the Council of Ministers, academia, domestic businesses such as Aramco, and the international business community. Yet, the true center of Aramco's analysis and policy formation, and one of the primary petroleum policy-setting groups in the kingdom, is the Corporate Planning Department, which new Vice President for Corporate Planning Abdulaziz Al Khayyal heads. The Planning Department prepares all relevant studies that Aramco senior management may require, including in-depth and highly technical analyses of the world energy situation, the international petroleum markets, and local geological information.

According to a senior executive in Elf Aquitaine's Middle East division, the Planning Department led the fight against allowing the IOCs into the kingdom's upstream energy sector in late 1998 and early 1999.[5] The department fought the plan not with political cajoling but with a barrage of compelling data, showing that, in the short- to mid-term, the plans were not economically sound.

Brief History

Although the U.S. oil companies that formed Aramco had originally agreed as part of their concession to train Saudi nationals for senior positions throughout the firm, Aramco remained for many years a U.S. outpost in the Saudi desert. In the 1950s, direct Saudi influence on the internal affairs of the company was so slight that Aramco had its own political and intelligence branches.[6] As late as the 1970s, most of the upper management and key engineers continued to be American.

In the early 1970s, the late King Faisal orchestrated the Saudi purchase of Aramco, through Oil Minister Zaki Yamani. Faisal had to ensure that enough Saudi nationals were qualified to assume daily control of the massive organization and also negotiate with the Americans not eager to relinquish what had been called "the richest commercial prize in the history of the planet."[7] What happened at Aramco was not unique; in the 1970s, oil-producing countries worldwide were nationalizing oil companies. Although Aramco was nationalized in 1976, its U.S. executives remained for more than ten years until 1988, when King Fahd officially established the Saudi Arabian Oil Company, or Saudi Aramco, by royal decree.

Despite skepticism by many Western observers, the company has thrived under domestic management. Royal Dutch Shell Chairman Mark Moody-Stuart admits that today "Saudi Aramco feels, smells, like a normal oil company."[8] As one sales executive from a Western technology company noted, "I've been selling to the state-owned oil companies, like Saudi Aramco and the UAE's Adnoc, for many years and they are always at the cutting edge of technology. We speak the same language. It is very easy for them to comprehend and absorb the latest technologies."[9] Finally, any doubt that the Saudis are capable of managing the complexities of the world's largest oil company would be dispelled with a visit to Shaybah—a $2.5 billion oil production facility (complete with a medical center, airport, recreation hall, etc.) conceived and operated primarily by Saudis.[10]

Saudi Aramco Board of Directors	
Eng. Ali Al Naimi	Minister of Petroleum and Mineral Resources
Dr. Ibrahim Al Assaf	Minister of Finance and National Economy
Ousama Faqih	Minister of Commerce
Dr. Musaed Al Ayban	Minister of State
Dr. Abdulaziz Al Dukhayyil	Rector, King Fahd University of Oil and Minerals
Abdallah Jumah	CEO and President, Saudi Aramco
Abdulaziz Al Hokail	Exec. VP for Manufacturing Operations
Dr. Sadad Husseini	Exec. VP for Exploration and Production
Abdallah Al Saif	Senior VP for Producing Operations
Rodney Wagner	Former Vice Chairman, Morgan Guaranty
Harold Haynes	Former Chairman, Chevron
James Kinnear	Former CEO and President, Texaco

Current Operations

Saudi Aramco currently manages all oil and natural gas activities in the kingdom and acts as the largest Saudi training ground for preparing technocrats, managers, and engineers to serve throughout the Saudi economy. Saudi Aramco's subsidiaries include Vela International Marine Ltd., Aramco Services Co. of Houston, Aramco Overseas Co. of the Netherlands, Saudi Petroleum International, and Saudi Petroleum Overseas. Vela International Marine owns and operates ships delivering crude oil to the United States, Rotterdam, and the Caribbean. Aramco Services and Aramco Overseas purchase drilling and other equipment. Saudi Petroleum International, located in New York, and Saudi Petroleum Overseas in London gather market and commercial data and provide recommendations and analysis for Saudi Aramco's international operations unit.

The majority of Saudi Aramco production is based in four major oil fields located in the Eastern Province of Saudi Arabia (see chart below); several smaller fields are also located in that province. All Saudi oil wells are free flowing at the well head, but advanced technology is necessary to optimize recovery and maintain field pressure. Since the 1990–91 Persian Gulf War, a more pressing concern for Saudi Aramco has been for diversification—that is, the exploration and development of areas that had until then not been touched. Although reserves are high, Saudi Aramco is developing more resources outside the Eastern Province, such as new fields in central Arabia. Currently, sustainable production capacity is officially 10.5 million barrels per day (bpd), although its surge capacity is well above that figure.

Saudi Arabia is currently developing four major gas projects, with the goal of increasing overall gas processing capacity from 4 billion cubic feet a day (bcf/d), which the integrated Master Gas System (MGS) achieved in 1998, to 6.6 bcf/d in 2002. These projects are the largest Saudi gas sector development in more than a decade and are taking place in response to increasing domestic demand for gas. The MGS provides fuel to the petrochemical and industrial complexes at Jubail on the east coast and Yanbu on the kingdom's west coast. It also powers utilities, supplies natural gas liquids for both domestic use and export, and provides liquid sulfur that is then palletized for export.

Major Oil Fields of Saudi Arabia			
Field Name	Proven Reserves (bn barrels)	Production Capacity (mn bpd)	Notes
Large Fields			
Ghawar	70–85	5.30–5.50	Largest on-shore oil field in the world.
Safaniyah	19–22	1.40–1.60	Largest off-shore oil field in the world.
Abqaiq	17–19	0.80–0.90	On-shore oil field.
Berri	10–12	0.65–0.80	Off-shore oil field.
Medium-to-Small Fields			
Manif	9–11	0.28–0.33	Off-shore oil field.
Zuluf	7 to 9	0.65–0.73	Off-shore oil field.
Shaybah	6.5 to 8	0.45–0.53	On-shore oil field recently developed.
Abu Safah	5 to 6.5	0.14	Off-shore oil field shared with Bahrain.

Sources: International Energy Agency, *Middle East Oil and Gas* (1995), pp. 188–190; author's personal calculations based on confidential interviews with U.S. government agencies and IOCs.

Sovereignty over the Divided Zone is shared by Saudi Arabia and Kuwait. Kuwait administers all oil production in its section of the zone, making the Divided Zone the only place where there is direct foreign involvement in the kingdom's upstream petroleum sector. It contains approximately 5 billion barrels of proven oil reserves, administered by Saudi Arabian Texaco and the Japanese-run Arabian Oil Company (AOC). Saudi Arabian Texaco operates the onshore fields in the Saudi portion of the Divided Zone, while AOC is a joint venture between Saudi Arabia (which holds 10 percent of the company), Kuwait (also 10 percent) and several Japanese companies (80 percent). The AOC holds concessions to the Divided Zone's offshore oil resources. (See page 59 for current negotiations surrounding the Divided Zone.)

Saudi Aramco operates all of Saudi Arabia's oil terminals and pipelines through its manufacturing operations unit. The main export terminals are located on the Arabian Gulf. The Ras Tanura terminal is the world's largest oil-loading port. A pipeline called Petroline carries crude oil from the east to refineries in the west and feeds the main export outlet on the Red Sea. Petroline is important because Saudi oil production is concentrated in the east, which is a more vulnerable area given its proximity to the Persian Gulf and the Strait of Hormuz. A Red Sea export outlet, therefore, is necessary to ensure smooth operations.

Saudi Aramco's manufacturing operations division manages nearly all the refineries in Saudi Arabia. Its primary function is to meet domestic market needs. Saudi Aramco is the sole domestic supplier, although retail service stations are independent. There are eight refineries in Saudi Arabia, five of which—Ras Tanura, Jeddah, Riyadh, Yanbu, and Rabigh— are fully owned by Saudi Aramco. Saudi Aramco and Mobil have equal ownership of the Yanbu-Samref refinery, and Saudi Aramco shares ownership of its Jubail-Sasref refinery with Royal Dutch/Shell. The AOC owns and operates the Khafji refinery, located in the Divided Zone.

Domestic Product Pricing

Economic problems in 1994 forced the government to cut energy-related subsidies, leading to a doubling of retail gasoline prices in Saudi Arabia to approximately $0.53 per gallon. In 1999, the price was raised again to $0.99 cents per gallon, a move the crown prince initiated.[11] At this price domestic gasoline prices are above cost and the government gains revenue from gasoline sales, unlike in Iran or Iraq where gasoline is so heavily subsidized that it costs less than $.05 a gallon—

an indication of the soundness of economic policy in Saudi Arabia compared to that of some of its neighbors. These and other related price increases lowered oil demand but only temporarily. Natural gas prices in Saudi Arabia rose by 50 percent in 1998 (from $0.50 to $0.75 per million British thermal units [BTUs]), the first increase in twenty years. Nevertheless, Saudi gas supplies are still among the most inexpensive in the world.

Foreign Refining Ventures

Saudi Aramco is involved in international partnerships with private oil companies in the United States, Greece, South Korea, and Philippines. The company has employed a strategy based on partial ownership stakes of 50 percent or less, gaining outlets without assuming extra administrative duties. Saudi Aramco's international partnerships provided outlets for 1.3 million bpd of crude oil, or approximately 20 percent of total exports, in 1998, and the company is currently establishing new partnerships in China and India.

Other Key Aramco Officials

Aside from Aramco president and CEO Abdallah Jumah, who was mentioned in chapter three in his capacity as member of the Supreme Council for Petroleum and Minerals Affairs, several other Aramco officials play important roles in the Saudi petroleum community and merit discussion. Among them are the following:

Abdulaziz Al Hokail, Executive Vice President of Manufacturing Operations. Abdulaziz Al Hokail joined Aramco in 1964 and was appointed to his current post in 1992. He is also a member of the Saudi Aramco board of directors. He has a bachelor's degree from the University of Texas, and he completed Carnegie Mellon University's Executive Program for mid-level managers in 1975. In his current position, his main responsibilities are oversight of all Saudi Aramco's refining operations. Al Hokail, according to a former senior official at Chevron with long standing ties to Aramco, is credited for being a dynamic manager and one of the architects

Key Officials in Saudi Aramco	
Mr. Abdallah Jumah	CEO and President
Mr. Abdulaziz Al Hokail	Exec. VP for Manufacturing Operations
Dr. Sadad Husseini	Exec. VP for Exploration and Producing
Mr. Abdallah Al Saif	Senior VP for Producing Operations
Mr. Ali Saleh	Senior VP for Refining and Distribution
Mr. Ali Seflan	Senior VP for Industrial and Internal Affairs
Mr. Saad Al Shaifan	Senior VP for International Operations
Mr. Dhaiffalah Al Utabi	Senior VP for Engineering Services
Mr. Mohammad Iraani	Senior VP for Financial Services
Dr. Ibrahim Al Mishari	VP for Information Technology
Mr. Abdulaziz Al Khayyal	VP for Corporate Planning
Mr. Yusuf Rafie	VP for Petroleum Engineering
Mr. Stanley McGinley	General Counsel

of Saudi Aramco's successful entry into the U.S. market through the Motiva partnership with Royal Dutch/Shell and Texaco.[12] He is also a leading proponent of consolidating the presence of the Saudi Arabian Oil Company in East Asia, especially in China. He is the de facto number two man at Saudi Aramco and the leading candidate to succeed Jumah as president and CEO when Jumah retires.

Sadad Husseini, Executive Vice President of Exploration and Production Operations. Sadad Husseini joined Saudi Aramco in 1972 and was appointed executive vice president for exploration and producing, with a seat on the board of directors, twenty years later. Husseini holds a Ph.D. in geology from Brown University and completed the Program of Management Development at Harvard University in 1982. He is known for his expertise in the technological aspects of petroleum production. In the mid-1980s, he was the main force behind Aramco's acquisition of its first supercomputer—at a time when only nations in the North Atlantic Treaty Organization (NATO) possessed them. Regarded as the third-highest-ranking person in the company, his non-Saudi origins pose an obstacle to him assuming a higher position.

Abdallah Al Saif, Senior Vice President of Producing Operations. Abdallah Al Saif joined Aramco in 1960 as a petroleum accounting clerk. Five years later, he was awarded a full Aramco scholarship to study abroad, resulting in a petroleum engineering degree from the University of Oklahoma in 1970. He completed the Management Program for Executives at the University of Pittsburgh in 1979 and was appointed to the Saudi Aramco board of directors on January 2, 1999.

During the 1980s and 1990s, he held various positions within the Aramco management bureaucracy, primarily in the Producing Operations Directorate. In 1982, he was awarded his first executive position as vice president for southern area producing, and he then assumed vice president positions in directorates such as Manufacturing, Supply and Transportation; Planning; and Sales and Marketing. In 1995, he became senior vice president for producing operations and was named an executive vice president in December 1998. He is a prime candidate to take over either of the two most strategic positions at Aramco: executive vice president for manufacturing operations, when Al Hokail moves into the top position; or executive vice president for exploration and producing, when Husseini retires.

Notes

1. *Saudi Aramco Dimensions Quarterly,* Winter 1998/1999, pp. 3–5; author's estimates from various sources.

2. See *Aramco,* May 1999, online at http://www.aramco.com/aboutus/orgchart.html.

3. Energy Intelligence Group (EIG), "Saudi Aramco," part of EIG's series *The World's Key National Oil Companies,* January 1999, p. 7.

4. Of this production, Iraq exported 2.6 million barrels and Kuwait exported 1.5 million barrels. M. A. Adelman, *The Genie Out of the Bottle: World Oil since 1970* (Cambridge, Mass.: MIT Press, 1995), p. 292.

5. Author interview in Paris, February 25, 1999.

6. Anthony Cave Brown, *Oil, God, and Gold: The Story of Aramco and Saudi Kings* (Boston: Houghton Mifflin, 1999), p. 129.

7. Ibid., p. 120.

8. Sarah Miller, "Royal Dutch/Shell's Moody-Stuart Tours The Global Oil Horizon," *Petroleum Intelligence Weekly,* November 9, 1998, p. 6.

9. "Ascend's Alhusan on Middle East Oil Producers," Bloomberg, May 27, 1999.

10. Michael Georgy, "Saudi Desert Offers Oil Investment Guide," Reuters, February 22, 1999.

11. Faiza Saleh Ambah, "Crown Prince Popular with Saudis," Associated Press, August 4, 1999.

12. Author interview in San Francisco, August 5, 1999.

Chapter 5

Response to Weak Oil Prices

In 1996 and 1997, prospects for the Saudi economy were bright, and relatively high oil prices had aided the Saudi budget. Yet, the first great test of the crown prince and his senior aides came with the drop in petroleum prices during 1998, and the subsequent drastic decline in the kingdom's oil revenues. The government reacted by instituting domestic austerity measures, working to bring U.S. oil companies back into the kingdom's energy sector, and orchestrating petroleum production cuts to raise oil prices. The combined impact of the austerity measures and higher oil prices is estimated to have reduced the 1999 government budget deficit by $6 billion, equivalent to 4.2 percent of gross domestic product (GDP). Nevertheless, continuing reforms will be needed for the kingdom to achieve its goal of eliminating the remaining $9 billion deficit, equivalent to 6.4 percent of GDP.

Economic Background: Price Drop and Recession of 1998

Although oil prices began their slow decline after 1982—except for a partial recovery during 1988–92—no one was prepared for the precipitous price plummet in 1998.[1] The price of benchmark West Texas Intermediate crude oil dropped from an average of $21 per barrel in 1996 to less than $10 per barrel in December 1998. This decline wreaked havoc on the economies of oil-producing nations. By the end of 1998, prices were down more than 30 percent for the year.[2] From January 1998 to March 1999, when oil prices began their recovery, weak oil prices had wiped out more than $82 billion from the economies of oil exporting countries.[3]

As a country that derives 90 percent of its export earnings from petroleum, the sharp drop in oil prices was disastrous for

Saudi Arabia. By the close of 1998, as the real price of oil dipped to 1973 levels, Saudi Arabia entered its first recession in six years. Official statistics show the country's economy contracted by almost 10 percent in 1998.[4] Average income was officially $7,222 in 1998 and was projected to fall below $7,000 for 1999, based on the official population figure of 18 million.[5] But the average income was most probably not as low as reported by the government and the press, though. According to a U.S. Embassy internal report, prepared with the help of the Central Intelligence Agency (CIA), a more accurate population estimate would be between 10 million and 12 million, making average income about a third higher than reported, between $10,800 and $13,000.[6] Regardless of which figure is accepted, per capita income has decreased substantially from the early 1980s, when it topped $28,000.

As serious as was the drop in income, the recession's worst impact was probably the exacerbation of an already serious social problem: unemployment. According to economist Ghazi Obaid-Madani, the rector of King Abdulaziz University, Saudi Arabia's unemployment figure for 1998–99 was around 27 percent.[7] In addition, 100,000 young men were entering the job market each year by the end of the 1990s.[8] Notably, construction in Saudi Arabia, a bellwether industry, also contracted by 10 percent in 1998.[9] According to Said al-Shaikh, the chief economist at the National Commercial Bank, "In the last year, our eyes have been opened to things we haven't seen before. There is no doubt these are major challenges for Saudi Arabia."[10]

In addition to cutting average income and exacerbating unemployment, the oil price decline also worsened the government budget deficit. The Saudi government had calculated the budget for fiscal year 1998 under the assumption that oil prices would remain at the $14.50 level. As prices fell far below that, the deficit grew dramatically, to an estimated $11.5 billion to $15 billion.[11] Saudi officials insisted that domestic borrowing could cover the bulk of this shortfall, but others were not sure Saudi banks could withstand this pressure indefinitely.[12] The kingdom's senior leadership did not try to conceal the magnitude of the problem. Statements by Crown Prince Abdullah at the December 1998 Gulf Cooperation Council (GCC) summit in Abu Dhabi summed up the situation well: "The days of the oil boom are over."[13]

Immediate Response: Cutting Expenditures and Raising Revenue

To deal with this sharp decline in revenue, the crown prince acted quickly with a series of measures during 1998 and early 1999 to cut government expenses. For example, all government ministries were ordered to cut spending by 10 percent.[14] Moreover, recruitment and wages in the government sector were frozen, and state contract payment deadlines were extended to their maximum length.[15] In addition, many specific projects were either canceled or substantially curtailed. Saudi Aramco suffered a major staff reduction, and outlays for a planned upgrade at the Rabigh refinery were cut by 60 percent.[16] Also, construction on the Haradh gas processing plant and work on three other gas-oil separating plants (GOSPs) were delayed.[17] Three dam projects (Wadi al-Laith, Rabigh, and al-Hali), intended to supply the parched capital with water, were also postponed.[18]

Following its weekly cabinet meeting on December 28, 1998, the Saudi government released its 1999 budget, which signaled that the cuts instituted in the previous year would continue. Specifically, the budget revealed a general willingness by the Saudi government to adapt its development prerogatives to the economic realities imposed by sinking oil prices. Expenditures were forecast at $44.6 billion and revenues at $32.7 billion (for a deficit of $11.9 billion). Social subsidies were reduced 50 percent from 1998 and defense expenditures by 25 percent to 30 percent (among the biggest casualties was a $1.7 billion arms deal with South Africa, which was shelved in February 1999).[19] There was no doubt that the priorities outlined in the new budget would have sizable effects on Saudi society and that the cuts revealed a strong commitment by the country's leadership to finally face its fiscal problems squarely.

In addition to these cuts, in 1999 the Saudis began to

institute several taxes and to raise prices; these efforts produced a tax on luxury goods, higher fees for obtaining a driver's license, an airport tax of $13 (excluding pilgrims but still expected to raise $240 million in 1999), and an increase in the permit fee for foreigners working in the kingdom.[20] This last move was meant not only to raise revenue but also to create more jobs for Saudi nationals and help staunch the annual outflow of $15 billion to $16 billion, repatriated by Saudi Arabia's 4 million expatriate workers.[21] As noted above, prices for gasoline were also raised by 50 percent during 1998 and 1999. Finally, at the end of 1998, the Saudi government warned mobile phone users to pay their delinquent bills or their service would be cut off.[22] These small moves were part of a larger effort to raise revenue and accustom the population to increased taxation. "It all adds up," said one economist. "There is a much better discipline on spending."[23] These efforts raised approximately $1.3 billion in new revenues, the equivalent of 1 percent of GDP.[24]

Return of the International Oil Companies (IOCs)

In addition to these domestic initiatives, the senior Saudi leadership began bringing the major U.S. and other international oil companies (IOCs) back into the kingdom. Significantly, the crown prince avoided serious internal opposition on this proposal by convincing the IOCs to submit proposals entirely on the kingdom's terms.

With population growth at 3.4 percent annually,[25] the kingdom's water, energy, and infrastructure needs rapidly expanded during the 1990s. In fact, if demand for electricity increases at a modest 4.5 percent annually, the government will have to raise an estimated $120 billion for power generation projects over the next twenty years.[26] Consequently, Saudi Arabia has much to gain economically from bringing foreign investment into the electricity and water sector. (In Saudi Arabia, the electricity sector includes the water sector, because water is desalinated when it is boiled in oil- or gas-fueled plants, producing both electricity and drinking water.) Although the state could collect additional funds for these projects through more borrowing—it could easily utilize Saudi Aramco's strong credit, a measure it employed in 1998 and which it is likely to use again—it has an aversion to incurring foreign debt. In any case, signs of strain—such as a water shortage in Jeddah in 1998—are becoming more common and are generally blamed on a lack of funds for investment in infrastructure.[27]

Speculation that the Saudis would bring the IOCs back into the kingdom was fueled by Kuwait opening ten of its fields to development bidding by Amoco, Arco, Chevron, Conoco, Mobil, and Texaco in November 1998, despite some Kuwaiti domestic objection to selling what the Kuwaitis call their "crown jewels."[28] In addition, Iran was moving forward with efforts to increase foreign investment in its own petroleum industries,[29] as were the Iraqis, who had negotiated contracts with Elf Aquitaine and Total–Petrofina that would become effective as soon as the United Nations lifted its sanctions.[30]

Those inclined to believe that the Saudis would follow suit were encouraged when the Saudi ambassador to the United States, Prince Bandar bin Sultan, invited seven U.S. oil executives to a meeting with Crown Prince Abdullah and Foreign Minister Prince Saud at Prince Bandar's Washington, D.C., residence. For some time, Prince Saud had been concerned about what he perceived as a lack of dynamism in the relationship between the United States and Saudi Arabia, and he started looking for ways to revitalize this relationship. This effort led to, among other things, the idea of inviting U.S. companies to increase their investment in the kingdom's energy sector. Before officially proposing a meeting, he vetted the idea with some current and former heads of such major U.S. oil companies as Texaco and Mobil, and they all expressed their enthusiasm at the possibility. Hence, after numerous discussions with senior officials in the Saudi government, it was agreed that such a conference would take place during the crown prince's trip to Washington in September 1998.[31]

At this gathering, the Saudis asked the U.S. chief executive officers (CEOs) to prepare proposals for petroleum projects in the kingdom that were "mutually beneficial." Although upstream development of the petroleum sector in

Saudi Arabia was not guaranteed, the possibility tantalized all of the executives. Describing the motives for this change in policy, the crown prince said in a late 1998 interview,

> The kingdom has no alternative but to diversify and develop the sources of the national income by expanding investment channels, opening doors for international companies, and offering them lucrative incentives . . . [Saudi Arabia] has all the potentials for attracting capital—ensuring its protection and stability.[32]

Over the next seven months, all the companies represented at the September meeting submitted their proposals.

Those who believe Saudi Arabia should open its doors to foreign investment in the upstream oil sector highlight several key benefits that this move would bring. First, it could strengthen the U.S.–Saudi relationship, which has proven effective at buttressing Saudi security. Second, the benefits of competition would certainly force Saudi Aramco to increase efficiency and profitability. Finally, as one commentator highlighted in March 1999, the kingdom has an interest in attracting some of the $80 billion to $90 billion per year that the world's largest oil companies currently spend on new oil exploration and development in the fields of its competitors.[33]

On the other hand, some believe the disadvantages of allowing foreign investment in the energy sector currently outweigh the advantages. This is the view held by Oil Minister Ali Al Naimi and senior executives at Saudi Aramco and articulated by Dr. Muhammed Al Saban, an economic advisor to Al Naimi. In a February 1999 interview, Al Saban argued that as long as the kingdom has substantial excess production capacity (enough to maintain market stability in the event of any sudden shortages), there is little reason to invest in upstream oil projects. Also, he noted that it is unlikely any IOC would be willing to allow its operations to sit idle in case of further necessary production decreases. According to Al Saban, "the value added from the entry of foreign companies is frequently overstated."[34] He would prefer that the responsibility to "perform better" be left to Aramco management, not to competition. Finally, he argued that attracting foreign capital will not reduce the funds available for other projects around the world, since oil investment is a function of oil price and political considerations alone.

Indeed, as prices hit recent historical lows in December 1998, the arguments of those opposed to opening the upstream oil sector gained more cogency. Then, when Saudi officials announced additional production cuts in March 1999, increasing Saudi spare production capacity to more than three million barrels per day (bpd)—more than the entire output of several Organization of Petroleum Exporting Countries (OPEC) members—the possibility of upstream foreign investment dimmed further. Yet, the U.S. firms did not lose hope and prepared a variety of proposals, including "integrated" gas proposals to provide desalinated water, electric power, or petrochemicals as an end product.[35]

In February 1999, U.S. energy secretary Bill Richardson went to the kingdom to inquire, among other things, on the status of opening the upstream oil sector to U.S. firms. As it became clear that the Saudis would not offer this liberalization for the foreseeable future, the U.S. companies were asked to accept involvement restricted to natural gas projects and downstream oil and petrochemical projects. To emphasize this point, Richardson was treated to a day-long journey to the kingdom's "Empty Quarter" in the Southeast region to see how Saudi Aramco engineers had developed the 14.3 billion barrel Shaybah oil field, which has a current capacity of 500,000 bpd, with little foreign expertise.[36] When asked why the upstream oil sector would remain closed, Al Naimi stated bluntly, "Reasonable people do reasonable things."[37]

With their dreams of access to upstream oil dashed, the IOCs pinned their hopes on upstream natural gas proposals. Many Saudis believe natural gas to be a much better area for foreign investment because one of the main priorities of the kingdom is to build power generation and water desalination plants powered by natural gas. Using natural gas could save 300,000 bpd of oil—the amount that the kingdom used for its power generation in 1999. The specified goal for Saudi natural gas development is to increase production from 4.0 billion cubic feet (bcf) to 6.3 bcf per day by 2001.[38] Most admit this plan would require at least some form of foreign investment. Of course, the major incentive for the IOCs to invest tens of billions of dollars in these projects is the hope for future contracts in the upstream oil sector. "Those who invest in developing the industrial base of Saudi Arabia," Al Naimi said, "will probably be the ones who will be invited when and if the upstream (oil sector) is available."[39] The stakes were so high—Saudi Arabia's proven oil reserves are worth more than $2.7 trillion at current prices—that, as one senior U.S. oil executive remarked, "Nobody could refuse."[40]

All the proposals that the IOCs presented focused on investments that would optimize Saudi revenues within a managed supply environment. Although Al Naimi appeared to welcome the IOCs optimization proposals at the February 1999 press conference with U.S. energy secretary Richardson—"They are exactly what we are looking for"—but once the proposals were examined by both an Oil Ministry and an Aramco technical team, the oil minister expressed his coolness toward the IOC bids. In May, he told reporters that, as far as he was concerned, the ideas could not be adopted, and by late July 1999, the influential Oil Ministry officially withheld its support from any of the proposals—for upstream oil *or* natural gas. The Oil Ministry officials argued that not only did Saudi Aramco have plans for projects similar to those proposed by the American companies, but also that they could implement them more cost effectively.[41] According to several American individuals involved in the meetings, the Oil Ministry officials provided compelling reasons why foreign investment should not be allowed.[42] Indeed, the Oil Ministry has been presenting plausible alternatives to investment and providing numbers that show the Saudis can develop the fields themselves. Among other things, privatizing the oil sector does not necessarily require turning over Saudi assets to foreign investors; Saudi businessmen themselves control several hundred billion dollars' worth of foreign assets and may choose to invest in the oil sector.[43]

But financing muscle and technical prowess are not the

only criteria the leadership is using to evaluate the bids. Even if the evaluators were correct concerning the Saudis' ability to develop the fields on their own—and that is far from certain as their arguments are clearly influenced by their own needs—their recommendation does not mean that the proposals will not be adopted. Considerations such as the strategic value of bringing the IOCs (especially the U.S. companies) back into the country, the liberation of capital for other projects, and the advantage of bringing increased competition to the Saudi energy sector will all play a role in the final decision as to whether to allow foreign investment. As Hassan Husseini, a former mid-level official at Saudi Aramco and managing director of the newly created Gulf Petro-Minerals, told delegates at the Seventh Annual Middle East Petroleum and Gas Conference in Bahrain in April 1999,

> The expansion of oil and gas investment opportunities in the kingdom will ultimately hinge on economic public policy issues, not technical prowess. Despite Saudi Aramco's technical excellence and record of consolidation and integration of the Saudi oil industry, it has not been able to assure the kingdom of sustained growth in revenues.[44]

Or, as one U.S. oil executive put it, "This thing is bigger than Aramco."[45]

Regardless of if or when the Saudi government opens the upstream oil sector to foreign investment, there are other opportunities in the kingdom for U.S. companies. For example, it is possible that a U.S. firm will be offered the opportunity to acquire Japan's concession in the off-shore "Divided Zone" shared by Saudi Arabia and Kuwait. This forty-year concession, which expired in February 2000, consists of an 80 percent stake in the Japanese-owned Arabian Oil Corporation (AOC). The remaining 20 percent is split between the Kuwaitis and the Saudis. The Japanese have balked at the two main conditions for renewal: increasing the amount of oil they buy from the kingdom (from 1 million bpd to 1.8 million bpd) and covering the entire cost of building a massive railway network linking the oil-rich eastern province with the rest of Saudi Arabia. When the Japanese politely refused on the grounds that the deal did not make economic sense, the Saudis started preparing for an eventual transfer.[46]

Whether the Saudis agree to embrace another foreign partner, and who that partner ultimately is, will be a significant test of the kingdom's mood regarding liberalization. The leadership may opt for an internationalist rather than a protectionist stance, in which case the United States would be the beneficiary of this shift. Both the Saudis and Kuwaitis would prefer to see a U.S. interest in the zone for geopolitical reasons, but in any case, the Kuwaitis have stated repeatedly that they will accept the Saudi choice. The lead Saudi official for the concession is none other than Prince Abdulaziz, the pro-U.S., pro–foreign investment deputy oil minister.

As of January 2000, the major American oil firms had not yet submitted formal proposals for the Divided Zone. But if and when the decision is made to offer the concession to an IOC, the American companies will most probably be slightly favored. Such a deal would have enormous potential value. The offshore fields managed by the Arabian Oil Company have a total output of 350,000 bpd (0.5 percent of the world's total daily supply). Of course, the portion of this that the U.S. companies see would depend on the terms that they negotiate with the Saudi Oil Ministry. But many believe that the Saudi government would be generous with the Americans— possibly sharing 15 percent to 20 percent of total production revenues above the cost of operations.

Other than potential revenue, such an arrangement would provide a perfect training ground for the major U.S. oil companies to prepare for the moment when they finally reenter the kingdom's upstream oil sector. This event could occur when Saudi production reaches its maximum output and additional investment is needed to match the need. Given that (1) demand for Saudi oil is expected to grow by approximately 6 million to 7.5 million bpd in the next decade,[47] (2) there is currently "only" 3 million bpd of excess capacity, and (3) 1 million bpd of Saudi crude exports currently translates into $4.5 billion per year, U.S. companies will have a role to play in the future.

In the next five years, therefore, continued foreign investment can be expected in the petrochemical industry and in power plants and desalination plants. This kind of investment will achieve the goal of the Saudi leadership, without their having to address the sensitive topic of opening the kingdom's upstream oil sector to the IOCs, something Saudi Arabia is loathe to do after taking control of its own petroleum resources more than twenty years ago. "Overall, the Saudis handled this situation quite well," said former U.S. ambassador to Saudi Arabia Walter Cutler.[48]

March 1999 Oil Production Cuts

The other major policy achievement that the crown prince and the leaders in the Saudi petroleum community could claim in 1999 was the agreement at the March OPEC meeting in Vienna to cut oil production. After the meeting, because of the Saudi-orchestrated cuts, oil prices quickly made a recovery back to what Saudis regard as the "sweet spot" of $17–$20 per barrel.

In the early months of 1999, the Saudis faced a dilemma in responding to the drastic drop in prices. On one hand, they could choose a policy of price support; on the other, they could produce more to increase market share. The former would force them back into the role of "swing producer" that they had forsaken. The latter would infuriate Iran and spell the end of OPEC.[49] The power of Saudi spare production capacity cannot be underestimated; as one Nigerian oil ministry official said, "You Saudis with your excess capacity could easily drive us—along with the Venezuelans, Mexicans, Norwegians, and all the non-Middle East producers—out of business."[50]

In a telling move, the Saudis chose to cut production and support prices. The motive behind the production cuts was clear—one of the weakest oil markets in years placed severe pressure on all exporters, especially Saudi Arabia. Yet, at the beginning of 1999, most analysts doubted that the main players—Iran and Saudi Arabia—could agree on who would suffer the brunt of the cuts. The Saudis had insisted that 8 million bpd represented an inviolable production floor for the kingdom. At the same time, the Iranians contended that their 1998

March 1999–March 2000 Petroleum Statistics

Country	Proven Reserves (billion barrels)	March 1999 Cuts (bpd)	Post-agreement Production	Domestic Consumption	Total Exports (million bpd)
Saudi Arabia	262	585,000 bpd	7.438 mbd	1.205 mbd	6.233
Russia	57	100,000 bpd	6.029 mbd	2.570 mbd	3.459
Norway	27	100,000 bpd	3.180 mbd	0.220 mbd	2.960
Venezuela	73	125,000 bpd	2.720 mbd	0.440 mbd	2.280
Iraq	113	-	2.600 mbd	0.350 mbd	2.250
Iran	90	264,000 bpd	3.359 mbd	1.255 mbd	2.104
UAE	98	157,000 bpd	2.000 mbd	0.350 mbd	1.843
Kuwait	97	144,000 bpd	1.836 mbd	0.160 mbd	1.676
Nigeria	23	148,000 bpd	1.885 mbd	0.220 mbd	1.665
Mexico	48	125,000 bpd	3.235 mbd	1.690 mbd	1.545
Libya	30	96,000 bpd	1.227 mbd	0.200 mbd	1.027
Qatar	3.7	47,000 bpd	0.593 mbd	0.020 mbd	0.573
Algeria	9.2	57,000 bpd	0.731 mbd	0.200 mbd	0.531
Indonesia	9.2	93,000 bpd	1.187 mbd	0.970 mbd	0.217
Kazakstan	20	-	-	-	-
Oman	5.2	63,000 bpd	0.807 mbd	-	-

Compiled from various sources, including Reuters, Bloomberg, and author's personal estimates based on confidential interviews.

quota assignment had been based on erroneous calculations and wanted a higher output cap based on their own figures.

Yet, the pain inflicted on the Saudi economy by the drastic drop in prices compelled policymakers in Riyadh to reevaluate their stance. In a series of high-level meetings that preceded the OPEC gathering, the Saudis agreed to renegotiate Iran's quota and stunned many analysts by offering to slash their own production by 585,000 bpd. This cut pushed Saudi daily output below the 8 million bpd mark. Al Naimi commented on the drop below this unspoken floor: "There is nothing magic about that number . . . it is not sacrosanct."[51] Although Al Naimi orchestrated the cuts, the crown prince and Foreign Minister Saud Al Faisal made the decision "based on clear economic reasoning."[52]

It soon became clear that the value of the agreement to the Saudis far outweighed the potential loss of market share and oil revenue. Excess oil stocks around the world were forecast to shrink by the third quarter. By the end of April 1999, oil futures had topped $18 for the first time since February 1998. Within a month, price increases erased 40 percent of the 1998 price decline. By summer, prices had recovered 90 percent of the previous year's drop. OPEC's achievement by June 1999 of 91 percent compliance with production quotas contrasted sharply with the widespread cheating on some OPEC agreements in earlier years.[53] When prices approached $20 per barrel in July, the kingdom's economic outlook improved markedly.[54] As one oil analyst noted in May 1999, "If oil prices stay up around this $18 or $19 a barrel level for West Texas Intermediate, then you see the impacts on the Saudi budget deficit [will be] quite significant this year. It will cut it in half, an anticipated $12 billion deficit would be down around $6 billion or less."[55]

In retrospect, the Saudi orchestration of the production cuts and the resulting hike in prices was a policy success. To begin with, the Saudis were able to raise much-needed revenue (what was gained from higher prices more than compensated for lower production). Moreover, they not only reinforced OPEC solidarity and provided a means for a breakthrough in relations with Iran, but also avoided the risk of annoying oil consumers, chief among them the United States, where the price increase went largely unnoticed in the context of a booming economy. At their May 1999 summit, Crown Prince Abdullah and Iranian president Muhammad Khatami announced in a joint statement that they had "reviewed the present situation in the oil market and expressed satisfaction with the outcome of the OPEC agreement of March."[56] Western analysts shared this satisfaction. "I was pleasantly surprised," U.S. deputy assistant secretary of state for Near Eastern Affairs Ronald Neumann stated, "by the tactical choices that the Saudis took in implementing their oil policy [in March 1999]."[57]

Notes

1. Jahangir Amuzegar, *Managing the Oil Wealth: OPEC's Windfalls and Pitfalls* (London: I.B. Tauris Publishers, 1999), p. 152.

2. "Consolidating for Growth," *Middle East Economic Digest (MEED)*, January 22, 1999, p. 7.

3. "Weak Oil Forced Gulf States to Cut Gas Spending," Reuters, May 19, 1999.

4. "Saudi Arabia Quarterly Report," *MEED*, January 31, 1999, p. 7.

5. See, for example, Saudi Arabian Monetary Agency's annual reports.

6. Peter W. Wilson and Douglas F. Graham, *Saudi Arabia: The Coming Storm* (New York: M.E. Sharpe, 1994), p. 12; U.S. estimates are generally viewed as free from Saudi political and national security pressures to inflate the population figure, and they are therefore considered more accurate.

7. See *al-Sharq al-Awsat*, June 14, 1999, p. 13.

8. Douglas Jehl, "Riyadh Journal: Rival Princes Ease Desert City's Horizon Skyward," *New York Times*, April 6, 1999, p. A4.

9. "Drop in Construction Hurts Saudi Cement Firms," *Reuters*, February 15, 1999.

10. Douglas Jehl, "For Ordinary Saudis, Days of Oil and Roses are Over," *New York Times*, March 20, 1999, p. A1.

11. *Energy Compass* 9, no. 46, p. 6 ($11.5 billion); *MEED* no. 27, November 1998, p. 21 ($13 billion).

12. "Saudi Arabia Quarterly Report," *MEED*, January 31, 1999, p. 7.

13. Faiza Saleh Ambah, "Saudi Crown Prince Exhibits Common Touch," *Washington Times*, August 11, 1999, p. A12. See also "Saudi Arabia: 'Text' of Prince 'Abdallah's 7 Dec GCC Speech," *al-Sharq al-Awsat*, December 8, 1998, p. 2; translated by Foreign Broadcast Information Service Near East and South Asia (FBIS-NES) Daily Report, December 21, 1998.

14. Toby Ash, "Trying to Make Ends Meet," *MEED*, November 27, 1998, p. 24.

15. Ibid.

16. Ibid.

17. Energy Intelligence Group (EIG), "Saudi Aramco," part of EIG's series *The World's Key National Oil Companies,* January 1999, p. 8.

18. "Low Oil Prices Delay Saudi Dam Projects," Reuters, February 14 ,1999.

19. Barry May, "First Victim of Saudi Cuts is $1.7 Bln Arms Deal," Reuters, February 9, 1999.

20. "Saudi to Introduce Departure Fee at Airports," Reuters, May 29, 1999.

21. Peter Kemp, "Breathing Space," *MEED,* June 11, 1999, p. 7.

22. Toby Ash, "Trying to Make Ends Meet," *MEED*, November 27, 1998, p. 22.

23. Diana Abdallah, "Saudi Seen Meeting Payments, May Issue More Bonds," Reuters, June 15, 1999.

24. For comparison, a tax increase of 1 percent of GDP in the United States would be $80 billion.

25. Jehl, "Riyadh Journal," p. A4.

26. "Saudi Invites Foreign Investment in Gas, Rules Out Oil," Bloomberg, February 7, 1999.

27. Ash, "Trying to Make Ends Meet."

28. *Energy Compass,* November 16, 1998, p. 1. In any case, the Kuwaitis had reasons other than economics for opening up their fields. A string of Western assets along the Kuwaiti-Iraqi border could be seen as a form of insurance against another attack by Kuwait's northern neighbor.

29. Michael Georgy, "Saudi Desert Offers Oil Investment Guide," Reuters, February 22, 1999.

30. Author interviews with oil company executives in Los Angeles, January 1999, and in London, March 1999.

31. Ibid.

32. Ash, "Trying to Make Ends Meet," p. 23.

33. Nathaniel Kern, "Oil Prices and Foreign Investment in Saudi Arabia," *Middle East Economic Survey* 52, no. 14, April 5, 1999.

34. Dr. Muhammed Al Saban, *Energy News Digest,* February 22, 1999, digested and translated from *al-Sharq al-Awsat,* February 20, 1999.

35. "U.S. Companies Must Lower Sights over Saudi Oil Reserves Access," Bloomberg, February 8, 1999.

36. The Shaybah complex, situated on top of 14.3 billion barrels of proven reserves, went onstream at the end of 1998 following a $2.5 billion investment. Philippe Rogier, "Price Development and Investment Opportunities: The Opening of the Middle East and North Africa,"

presentation at the Center for Global Energy Studies, Ninth Annual Conference, London, April 1999.

37. "Saudi Invites Foreign Investment in Gas, Rules Out Oil," Bloomberg, February 7, 1999.

38. EIG, "Saudi Aramco," p. 9.

39. Amir N. Ghazar, "Middle Eastern Nations Gear Up for Massive Petrochemical Expansions," *Chemical Market Reporter,* March 12, 1999, p. 7.

40. Author interview in Los Angeles, January 14, 1999.

41. "Saudi Aramco Turns Down First Upstream Offers," *Petroleum Intelligence Weekly* 38, no. 21 (May 24, 1999), p. 1.

42. Author interview with American Aramco managers in Baltimore, May 14, 1999.

43. Currently, there are two privately held domestic oil companies in the kingdom—Nimr Petroleum and Delta Petroleum—owned by peripheral members of the royal family.

44. Toby Odone and Jareer Elass, "Saudi Arabia: Public vs. Private," *Energy Compass*, May 21, 1999, p. 4.

45. Georgy, "Saudi Desert Offers Oil Investment Guide."

46. "Saudi in Proceedings for Neutral Zone Takeover-MEES," Reuters, June 21, 1999.

47. These figures are based on the Energy Information Administration's "Global Energy Demand Forecasts, 1999," published by the U.S. Department of Energy.

48. Author interview in Washington, May 27, 1999.

49. "Saudi Arabia Quarterly Report" *MEED*, January 31, 1999, p. 22.

50. Author interview in Vienna, March 19, 1999.

51. Georgy, "Saudi Desert Offers Oil Investment Guide."

52. Karen Matusic, "Saudi to Alter Cabinet for Second Time in 25 Years," Reuters, June 14, 1999.

53. "OPEC Made 91 Percent of Promised Oil Output Cuts in May," Bloomberg, June 2, 1999.

54. "Crude Rises to 19-Month High; OPEC Seen Making Cuts," Bloomberg, July 6, 1999.

55. "Oil Prices Could Halve Saudi Deficit—Bank," Reuters, May 13, 1999 (quoting Brad Bourland, chief economist at the Saudi American Bank).

56. "Saudis, Iran Say They're Satisfied with Oil Price Increases," Bloomberg, May 19, 1999.

57. Author interview in Washington, May 21, 1999.

Chapter 6

Reducing Dependence
on Oil Income

D espite its achievements, even in the late 1990s the Saudi economy suffered from major structural problems. Massive subsidies, inefficient industries, and the lack of a fully developed free-market system all hampered the ability of the state to pull itself out of debt, which, according to a study by the Riyadh-based Consulting Center for Finance and Investment, reached 98 percent of gross domestic product (GDP) in 1999.[1] The root cause of the persistent budget deficit lies in the fact that growth in the domestic economy leads to an increase in expenditures without a concomitant increase in revenues, as would be the case if the government imposed taxes.

Apparently, senior leaders have realized that even the return of higher petroleum prices is not a permanent solution. The kingdom has had economic troubles even when oil prices have been relatively strong; budget deficits have plagued the state since 1982, with an average $2.4 billion annual deficit from 1986 to 1994.[2] As a U.S. Embassy report revealed, "The recent oil price recovery will only give the government some breathing space and will not be enough to seriously brighten prospects in 1999."[3] In any case, the higher oil prices that the production cuts created will be extremely difficult to maintain. In addition to the ever-present difficulties in maintaining a cartel—that is, the Organization of Petroleum Exporting Countries (OPEC)—the kingdom must hope that world demand will grow. Continued stagnation in Asia, another warm winter in the West, or the appearance of an overdue recession in the United States could quickly renew downward pressure on prices. Finally, new reserves; new international treaties on greenhouse gas emissions; new energy-creating technologies; or new exploration, extraction, or refining technologies could all force prices down in the long term.

Both the World Bank and the International Monetary Fund (IMF) have recommended structural changes in the Saudi economy—changes that are imperative if the kingdom is to attract private-sector investment and achieve its long-term goals of industrial diversification and elimination of the budget deficit. Topping the list of IMF recommendations is accelerating the drive toward economic reform by liberalizing foreign-investment laws and relaxing restrictions on non–Gulf Cooperation Council (GCC) nationals investing in Saudi companies; reforming the financial sector; and creating a comprehensive privatization program that details the government's strategy to disengage from commercial activities. As one Saudi banker said in 1994,

> Really, the solution is right in front of it. If the government cut back on its defense spending and all the payoffs that go with it, and sold its basic services even at cost—never mind at a profit—its deficits would disappear overnight. It should also think about stopping subsidizing foodstuffs, gasoline, and the airline. And why own the Intercontinental Hotel and the airline and the electric companies?[4]

The approach of the crown prince and the senior princes to solving the country's financial woes seems to be implementing significant reforms but at a measured pace. Key elements of their program include stepping up the process of privatizing key state industries, becoming more open to the idea of foreign investment, and working to further diversify the economy. The challenge will be implementing these reforms in the face of the pain that change will inflict on some sectors. In addition, the reforms need to occur at a rapid pace, given the severity of the problems facing the country.

Tapping the Private Sector

Few in Saudi Arabia perceive the same urgency for privatization as do foreign economists. Moreover, most Saudi policymakers support privatization in principle, but they have real qualms about each specific case. Thus, it may take years to resolve objections to privatizing existing state firms. Although Saudi Arabia is starting the process later than the non–oil-producing Middle Eastern states, those states—including Egypt, Jordan, and Israel—talked for years about privatization without actually doing much. Having avoided much of the debate, Saudi Arabia is actually at the same stage of privatization as the other states. Despite these objections, a true opportunity for progress currently exists, primarily because the crown prince himself has spoken out in favor of privatization.

Different privatization models must be examined to better understand what might be done in various parts of the oil industry. Khalid Al Zamil, president of the Saudi Council of Chambers of Commerce and Industry in 1999, made clear the kingdom's intentions when he remarked, "We are ambitious and want to see some of the oil and gas industries, such as refining and exploration, owned 100 percent by the private sector over the next five years."[5] Although the economic recession of 1998 brought the question of private investment in the oil sector to the fore, this trend has been underway for many years outside of Saudi Arabia. In fact, state oil companies around the world have been undergoing massive restructuring for the past decade, and the number of state oil companies (and the percent of state ownership of partially owned companies) has been declining steadily for more than ten years.[6]

Privatization had been discussed for years in the kingdom, beginning with the Fourth Development Plan of 1985–90. This plan broached the idea of privatizing the state's two largest firms, Petromin (which is now defunct) and the state airline Saudia, followed by the Saudi Consolidated Electric Companies (SCECOs). Among the many compelling reasons the plan listed for embarking on such a road was the potential repatriation of some $450 billion in private funds held outside the country.[7] Yet, little movement in this direction took place before the mid-1990s. King Fahd—in a commencement address at King Saud University in Riyadh—revitalized the notion in 1994 when he signaled his support for the idea:

> The state intends—and I have expressed this to the Council of Ministers—to relinquish many of the productive and beneficial services so that national capital can participate in them. . . . I believe that this is a contribution by the state to include citizens in constructive, beneficial, and useful services.[8]

The next year, the kingdom's sixth five-year economic plan (1995–2000) outlined the reasons for privatization. This document argued that privatization would be useful

- to achieve a balanced budget by the year 2000;
- to reduce the burden on the government treasury created by a large public-sector salary bill;
- to improve operational efficiency of state-owned firms; and
- to use privatization income to retire existing domestic government debt, interest payments of which then constituted 10 percent of country's annual budget.[9]

Once again, although movement was slow at first, the push to increase privatization gained steam after the confluence of events in the early- to mid-1990s, including the kingdom's Persian Gulf War debt and its continuing budgetary problems, which exacerbated declining oil revenues. Other factors included ongoing attempts to become a member of the World Trade Organization (WTO), which would require swift progress on economic liberalization, and a general anxiety over the need to employ the hundreds of thousands of Saudis expected to enter the job market each year.[10] Finally, the Saudi leadership was becoming more and more wary of relying on a single product (petroleum) for such a large portion of state income and wished to stimulate a diversification of the economy. These facts, coupled with a tightened fiscal policy aimed at reducing repeated budget deficits and an unwillingness to employ direct taxation to raise revenue, made strengthening the private sector particularly appealing. As one Saudi analyst noted, "By privatization, they will be able to repay at least some of the debt. Some of the subsidy will be taken out of the government budget and it will help them to invest in health and education."[11] The Saudi Electric Company, Saudi Airlines, Saudi Telecom, and the Saline Water Conversion Corporation have all been cited as potential candidates for privatization, although as yet none has been fully privatized.

The king tasked the crown prince and his closest advisers with the mission of privatizing key state industries and sectors. Like all major decisions in the kingdom, this one was made slowly and only after much debate and deliberation. Nonetheless, work progressed steadily toward this goal as the decade wore on. By 1998, according to Abdallah Dabbagh, former secretary general of the Saudi Chambers of Commerce and Industry and current member of the Consultative Council, the senior leadership had all agreed upon the necessity of privatizing the public utilities and services sector.[12] Both Crown Prince Abdullah and Prince Sultan, the minister of defense and civil aviation, expressed on numerous occasions their support for privatization. Also, the Council of Ministers approved the establishment of independent bodies to begin the privatization of state-owned enterprises. According to Brad Bourland, chief economist at the Saudi American Bank, "Economic reform issues in the kingdom do not rise and fall with the fluctuation in oil prices. There is a very real and broad consensus in Saudi to attract foreign direct investment regardless of [the] price of oil."[13] By 1999, the private sector had expanded to 35 percent of the kingdom's gross domestic product (GDP).[14]

The Saudi Consolidated Electric Companies

The electric power sector is perhaps the most important component of the economy that the government intends to privatize. Burdened by the obligation to sell power below cost, the four regional SCECOs, along with the six companies that provide power to the northern regions, have long been unprofitable. Although government payments keep the companies from bankruptcy, analysts point out that this system prevents company managers from conducting any sort of financial forecasting or planning.[15] Further, rapid population growth and increased demand for power (growing at 5 percent to 7 percent annually) has strained the current grid, requiring massive expansion to meet the kingdom's needs.

To address these problems, the crown prince and Prince Sultan instructed Electricity and Industry Minister Hashem Yamani to unify the SCECOs to become the Saudi Electric Company.[16] On November 30, 1998, the cabinet approved merging the ten power providers into a new vertically organized firm worth $8.9 billion and creating distinct entities for power generation, transmission, and distribution.[17] Following this reorganization, the government announced an electricity price increase of 20 percent for industrial users and some households.[18] Yet, to implement the government's new infrastructure plan fully, estimates indicate that the government would have to install 2,000 megawatts of new capacity each year through 2020, at a total cost of $120 billion, far too much for the government to raise by itself.[19] Saudi leaders thus realized that private investment would be an essential component of the scheme.

Saudi officials have considered a variety of private finance initiatives with regard to ownership, management, and financing.[20] Exactly how the government will decide to bring private firms into the power and electric sector is not certain. Simple international commercial loans are possible, but because of the debt of the current SCECOs, it would be difficult to utilize this avenue for all funding. The government will likely apply the successful strategies used when previously raising funds for energy-sector projects. For example, a special tax on heavy users of electricity funded the 1,200-megawatt power generation plant outside Riyadh.[21] Also, an internationally syndicated, $500-million loan (the first of its kind in the kingdom) provided funds for the Ghazlan power plant expansion, and the Saudi government employed joint ventures to upgrade utilities in Jubail and Yanbu.[22]

In any case, full privatization of the power companies will force the government to implement some difficult policies. It must ultimately remove subsidies—a move that is not politically appealing—or the government will have to continue to pay the difference between the market price and the subsidized price. In addition, outstanding power company debts—particularly for fuel bills from Aramco—will have to be settled. Finally, the government must address the issue of the power company's sectors in remote parts of the kingdom; because these remote parts of the country are more expensive to power, and are thus unprofitable, the government will likely have to make annual payments to a private firm to ensure that it provides power there at an acceptable cost. Nonetheless, the government seems determined to privatize the power industry substantially. The necessary steps to proceed with this long-awaited move have been taken; the question now is, at what pace will the government follow through with the most politically difficult step: bringing in the private sector?[23]

Saudi Arabian Airlines

With billions of dollars in assets and an extensive international flight schedule, Saudi Arabian Airlines is one of the kingdom's most powerful and prized enterprises.[24] It is the largest airline not only in the Middle East but also in the entire Islamic world. It recently upgraded its 113 large- and wide-bodied jet aircraft, which together carry about 12.5 million passengers and 225,000 tons of cargo annually. In 1998, the airline served fifty-two international and twenty-five domestic destinations, with a budget estimated at $2.26 billion.[25]

Yet, because of inefficiencies throughout the organization, lower-than-cost domestic airfares, and the custom of providing free tickets to government officials, the company has suffered from perennial losses. For example, in the otherwise austere 1999 budget, the Saudi government allocated more than $2.5 billion to support the company.[26] The Saudi American Bank estimated that in 1996 the company had a negative financial worth of $750 million, excluding physical infrastructure (e.g., airplanes, hangars, and facilities).[27]

The Saudi leadership decided in 1995 that, before proceeding with any privatization plans, it must first implement a planned overhaul of the company. Thus, soon after King Fahd's remarks about privatization, the airline's leadership implemented a modernization campaign. One of the first steps in this campaign was the purchase of sixty-one new aircraft from Boeing and McDonnell-Douglas (B747-400s and B777-200s), to be delivered over a five-year period. This purchase made Saudi Arabian Airlines one of the most well-equipped airlines in the world. But it also led many observers to believe that the ultimate privatization of the company was inevitable. The high cost of maintaining such a large fleet contradicted the government's avowed intention of eliminating the budget deficit; Saudi bankers and airline leaders therefore interpreted the purchase as signaling that the political leadership wanted the expense to be covered by a profitable, privatized company, not by the government. "The new fleet is the first step towards privatization," explained the airline's director-general, Khaled A. Ben-Bakr. "In fact, the new aircraft and privatization are two sides of the same coin."[28]

The government eventually decided to look to foreign banks for a loan to fund the aircraft purchase, rather than privatize before implementing the planned efficiency-enhancing program. In late 1997, Riyadh announced it was looking for $4.5 billion from international lending sources to pay for the U.S. planes, and it secured a syndicated international and domestic loan of $4.3 billion in 1998, guaranteed by the Ministry of Finance.[29]

Once the plane financing was complete, the government announced a "top-to-bottom" reform of Saudi Arabian Airlines to generate some profits and make it more attractive to private investors.[30] Ben-Bakr announced, "We are cutting costs, boosting productivity, and making ourselves more responsive to the needs of the customer."[31] Finally, the Saudi leadership squarely addressed the politically sensitive topic of raising prices: "We are making progress on prices. International fares have become more competitive. . . . Prices have moved closer to other airlines using the same routes."[32] Remarkably, the government has done little to force its citizens to use the airline. Unlike in other countries, there are no foreign-exchange restrictions on foreign airline tickets.

Saudi Arabian Airlines entered direct consultations with the Saudi private sector in 1995. As Ben-Bakr commented, "We are sitting together with businessmen . . . listening to their complaints, demands, and perceptions of Saudia [as the airline was then called]. The result is a clear picture of what they think of us and what we must do to improve."[33] The culmination of these meetings was the appointment in late 1995 of a new board of directors for the airline that included three Saudi businessmen. In addition, the new Board of Directors established a joint committee consisting of the airline board and other local businessmen to encourage the consultative process.

Another element of the airline reform process has been cutting labor costs, which in most countries would mean mass layoffs. The situation is more complex in Saudi Arabia, though. Certainly, many state firms are overstaffed, but most of the staff is foreign. The challenge is finding ways to hire more Saudis and pay them a salary justified by the work they perform, while still cutting costs. In some areas, this approach may be quite practical; for example, 95 percent of the airline's pilots are natives, and they may cost less than foreign pilots. In other areas, however, it may be more difficult to get value-for-money from "Saudization," and it will be difficult to avoid political pressures to hire more Saudis. For example, in 1996, the airline hired more than 3,000 Saudis for secretarial positions, an unusually large number to hire in one year given that the company has 56,000 employees.[34]

The outlook for the airline improved in areas beyond those mentioned above. Free ticket allowances to government officials and members of the royal family decreased significantly.[35] Government subsidies were still substantial, but they decreased from more than $3 billion in the early 1990s to $2.5 billion in 1998. In fact, the Saudi record on privatization of the national airline does not suffer by international comparisons. Across Europe—in France, Germany, and Italy, among others— privatization took many, many years of preparation, and Middle Eastern states have lagged even further, as evidenced by the decade-long debates about privatizing smaller airlines such as

Royal Jordanian, Egypt Air, or El Al in Israel. Thus, although it is true that there is little prospect that private investors will be able to buy part of Saudi Arabian Airlines in the near future, it is also true that the government is proceeding apace with the privatization process.

The Saudi Telecommunications Company

Strains on the country's telecommunications network have led to talk of privatization, which the crown prince strongly supports.[36] As a first step, the kingdom is "corporatizing" the telephone system to prepare it for the first Saudi privatization of a government utility. This process entails transferring its operations from the Ministry of Posts, Telegraphs, and Telephones (PTT) to the Saudi Telecommunications Company (STC). The STC was established in 1998 with an initial capitalization of $2.6 billion.[37] The U.S. firm J. P. Morgan was chosen to guide the company through this process.[38] The plan, endorsed by then–PTT Minister Ali Al Jehani, calls for the sale of shares to the public by the first or second quarter of 2000. Ultimately, Al Jehani has said he "would like to see it completely privatized and the ministry of PTT gone."[39] After that, he indicated that he would quickly move to allow competition for the new company, especially in cellular and data transmission.

In the meantime, the utility has been contracting out projects to expand the country's telephone network. In 1994, AT&T won the $4-billion contract for TEP-6—a project to expand telecommunications capacity by 1.5 million lines—and then subcontracted out the project to Lucent Technologies.[40] In 1998, the government announced the $700-million TEP-8 project to triple the kingdom's 2.4 million telephone lines, and by 1999 Lucent, Alcatel, Siemens, Ericsson, and Northern Telecom had submitted bids.[41]

The mode of PTT's privatization is noteworthy because Al Jehani seemed intent on removing any possibility of corruption or favoritism from the process. Although TEP-6 was awarded to a single company, Al Jehani insisted that future projects would be divided into smaller subcontracts. He also repeatedly reminded the companies that technical merit alone will decide who gets the business.[42] As a result, the lobbying of French foreign trade minister Jacques Dondoux, who arrived in April 1999 to promote the Alcatel bid, will probably not bear fruit.

Other Privatization Projects

In 1997, for the first time, the government gave a local business the contract to manage a reexport zone in the Jeddah Islamic port. In 1998, a similar contract for the King Abdulaziz port in Dammam was awarded to another private company.[43] Rumors further indicate the government will soon privatize the Saline Water Conversion Corporation—the agency that supervises all the large desalination plants in the kingdom and consumed $640 million of the 1999 budget.[44] Perhaps most important, systematic preparations are being created to develop the stock market, already one of the two largest in the Middle East, for increased foreign investment possibilities. The government in 1997 approved the first fund for investing in local firms that will be open to non–Persian Gulf nationals.[45] The Saudi American Bank (SAMBA) launched this Saudi Arabian Investment Fund, which is listed on the London Stock Exchange. Totaling only $500 million, the fund will make little difference in the estimated $48 billion Saudi market, but it is an important step in economic reform and possible privatization plans.[46]

Notes

1. "Study Says Saudi Government Debt $130.4 Bln—Paper," Reuters, May 24, 1999.

2. "U.S. Companies Must Lower Sights over Saudi Oil Reserves Access," Bloomberg, February 8, 1999. For annual deficit figures, see Jahangir Amuzegar, *Managing the Oil Wealth: OPEC's Windfalls and Pitfalls* (London: New York: I.B. Tauris, 1994), p. 155.

3. "Saudi 1999 Economy to Make Minor Gains—U.S. Report," Reuters, May 11, 1999.

4. Michael Field, "The Cheque's Not in the Post," *Euromoney*, May 1994, p. 101.

5. Kinda Jayoush, "Saudi Businesses Want Private Oil Industry," Reuters, May 15, 1999.

6. Energy Intelligence Group (EIG), "Saudi Aramco," part of EIG's series *The World's Key National Oil Companies*, January 1999, p. 5.

7. Abdallah Al Dabbagh, former secretary general of the Saudi Chambers of Commerce and Industry and current member of the Consultative Council, interview with author in London, April 12, 1999.

8. Edmund O'Sullivan, "Saudi Arabia: Prophet of Privatisation Makes a Mark," *Middle East Economic Digest (MEED)*, May 20, 1994, p. 26.

9. Moin A. Siddiqi, "Unleashing the Potential of the Private Sector," *Middle East* no. 273, December 1997, p. 29.

10. "Saudi 1999 Economy to Make Minor Gains—U.S. Report," Reuters, May 11, 1999

11. Peter Kemp, "Breathing Space," *MEED*, June 11, 1999, p. 7.

12. Author interview in Washington, May 1, 1999.

13. See Bloomberg, August 22, 1999.

14. Kemp, "Breathing Space."

15. Toby Ash, "Trying to Make Ends Meet," *MEED*, November 27, 1998, p. 28.

16. Ibid.

17. "Merged Saudi Power Firm Valued at $6.67 Bln.—Paper," Reuters, February 10, 1999.

18. Robin Allen et al., "Running on Empty," *Financial Times*, March 29, 1999, p. 15.

19. *U.S. Energy Administration Saudi Arabia Report, 1999*, May 15, 1999, located online at http://www.eia.doe.gov/emeu/cabs/saudi.html.

20. Robert Trevelyn, "Special Report: Saudi Arabia," *MEED*, June 13, 1997, p. 34.

21. *U.S. Energy Administration Saudi Arabia Report, 1999*.

22. John Cooper, "Getting into Shape for the Private Sector," *MEED*, April 5, 1996, p. 29.

23. Trevelyn, "Special Report: Saudi Arabia," p. 34.

24. Author interview with Ambassador Ahmad Abdel-Jabbar, former Saudi permanent representative to the United Nations (Geneva), in Geneva, February 20, 1999.

25. See the Saudi Arabian Airlines Website, http://www.saudiairlines.com/.

26. Although Saudi Arabian Airlines income is not precisely known, losses were estimated at over $1 billion. See Ministry of Finance and National Economy, "1999 National Budget Statement," released by the Saudi Press Agency.

27. David Pike, "Kingdom Line Flies through Tough Times," *MEED*, June 14, 1991, p. 13.

28. Pike, "Kingdom Line Flies through Tough Times," p. 14.

29. Kevin Taecker, "Update on the Saudi Economy, March 1999," *Saudi American Bank*, located online at http://www.us-saudi-business.org/tmar99.htm.

30. Cooper, "Getting into Shape for the Private Sector," p. 612.

31. Pike, "Kingdom Line Flies through Tough Times," p. 11.

32. Ibid.

33. Ibid., p. 19.

34. "Saudi Changes Image: Awaits Key Deal," United Press International, June 16, 1995.

35. Author interview with a member of the Saudi Consultative Council, in Geneva, April 5, 1999.

36. Faiza Saleh Ambah, "Crown Prince Popular with Saudis," Associated Press, August 4, 1999.

37. "Foreign Firms Still Await Word on Saudi Phone Deal," Reuters, June 29, 1999.

38. "Saudi Arabia Quarterly Report," *MEED*, January 31, 1999, p. 35.

39. Kemp, "Breathing Space," p. 15; Jehani was removed from his position in January 2000 and made a minister of state without portfolio.

40. Ibid.

41. "Saudi Arabia Invites Bids for One Million GSM Lines," Reuters, August 30, 1999.

42. Kemp, "Breathing Space," p. 15.

43. "Local Firm Wins Deal to Run Saudi Re-Export Zone," Reuters, February 21, 1999.

44. "Saudi Arabia Quarterly Report," *MEED*, January 31, 1999, p. 10.

45. "Samba Establishes First Vehicle for Foreign Investment in Saudi Arabian Shares," *Middle East Economic Survey* 40, no. 13 (March 1997).

46. Trevelyn, "Special Report: Saudi Arabia," p. 25.

Chapter 7
Saudi International Oil Policy: Relations with Exporters

S audi Arabia's foreign policy is in many ways a subset of its oil policy. As such, it has two distinct branches: relations with fellow exporters and those with oil-consuming countries. One of the main characteristics of the new leadership's common policy vis-à-vis both groups is its greater assertiveness in achieving Saudi policy goals, with bold moves or a subtle diplomacy that achieves the same ends.

Whereas many elements—economic, historical, religious, and political—make up the kingdom's relations with other oil exporters, those relations are heavily conditioned by one key set of facts: The kingdom possesses the largest petroleum reserves, the greatest spare production capacity, and the cheapest uplift costs in the world. As a result, Saudi Arabia has the ability to flood the market with oil, drive out most other high- and medium-cost producers, and ultimately become the world's preeminent supplier. Because this strategy has great political and economic risks, it has never been adopted. Yet, the very possibility of its implementation hangs likes a sword of Damocles over producers such as Mexico, Venezuela, Russia, and Norway, whose production costs are more than triple that of Saudi Arabia. Even other low-cost producers, such as Iran, the United Arab Emirates (UAE), and Kuwait, would suffer severe dislocations in the short to medium term before increased market share compensated for decreased prices. Because of this potential threat, all other oil-producing states pay heed to the Saudis.

Islamic Republic of Iran

Saudi Arabia and Iran have had hostile relations since the Iranian Islamic Revolution of 1979 when the Ayatollah Ruhollah Khomeini threatened to seize Saudi Arabia and return the holy cities of Mecca and Medina to the Shi'is.[1] The Saudis responded by lending the Iraqis nearly $40 billion for their war against Iran. The volatility of the Saudi–Iranian relationship continued eight years later when 400 people, mostly Iranian pilgrims, were killed in Mecca in clashes with Saudi security forces at an anti-U.S. demonstration.

The warming of relations between the kingdom and Iran could not have occurred without the election of Muhammad Khatami to Iran's presidency in May 1997. Once he took office in August in 1997, Khatami spoke of the desirability of improving relations with Arab Gulf states.[2] Crown Prince Abdullah took the first step by traveling to Tehran in December 1997 for the Organization of the Islamic Conference summit. In 1998 the two countries began transforming their relationship from adversarial to cooperative; for example, former Iranian president Ali Akbar Hashemi Rafsanjani visited the kingdom for two weeks that February. By 1999, the Saudi minister of defense had visited Tehran and President Khatami had visited Riyadh. These high-level meetings begat a string of press conferences at which leaders from both states lauded the new friendship and proclaimed a variety of cultural and economic exchanges. The Iranians pushed for military agreements, but the Saudis were not as enthusiastic about such proposals. Smaller improvements emerged throughout 1999 as Saudi Arabian Airlines agreed to restart flights to Iran after a twenty-year hiatus,[3] and the Saudis appointed a Shi'i—Jameel Al Jishi—as Saudi ambassador to Iran. This rapprochement was one of the most significant geopolitical changes in the Persian Gulf arena in years, and it was deeply influenced by the kingdom's oil policy.

The Saudis' need to address decreasing oil prices in a concerted way in 1998 converged with their desire to strengthen the Khatami government against Iranian hardliners. Accordingly, despite accusations that the Iranians had "flouted" a 1998

Middle Eastern States' Average Oil Production, 1994–98					
(in thousands of barrels per day)					
	1994	1995	1996	1997	1998
Algeria	755	767	820	850	816
Egypt	870	870	850	840	825
Iran	3,600	3,600	3,600	3,600	3,600
Iraq	515	545	590	1,270	2,151
Kuwait	2,105	2,095	2,155	2,000	2,088
Libya	1,415	1,420	1,440	1,390	1,378
Oman	870	820	880	895	860
Qatar	460	450	475	650	660
Saudi Arabia	8,100	8,100	8,200	8,600	8,393
Syria	610	565	605	600	600
Tunisia	90	95	90	80	80
UAE	2,100	2,100	2,200	2,290	2,273
Yemen	335	335	370	375	375
Other ME states	55	55	50	50	50
Total	**21,880**	**21,817**	**22,325**	**23,490**	**24,149**

Sources: BP Amoco, *Middle East Economic Digest, Petroleum Intelligence Weekly*, and author's personal calculations.

production agreement by implementing only 1 percent of promised cutbacks, the Saudis built on the growing contacts and trust they had created with the Iranians to forge a deal that would cut supplies and raise prices.[4] The deal was implemented at the March 1999 Organization of Petroleum Exporting Countries (OPEC) meeting in Vienna, and within months, prices approached $20 per barrel, nearly double what they had been in December 1998.

The March 1999 petroleum production cuts shed light on the regional focus of the Saudi leadership. To move forward on a general cutback, the Saudis had to accept the bulk of the cuts themselves. The concession to the Khatami government set the stage for Khatami's May 1999 visit to Riyadh, the first by an Iranian president in twenty years. Only with Iran safely subdued as a partner could the kingdom hope to make progress on other important regional issues.

Although Saudi Arabia agreed to bear the brunt of the production cuts (25 percent) and some perceived Iran as a winner, the high-level negotiations also forced Iran to share in the cuts, which it had failed to do in previous agreements negotiated in 1998. During the end of 1998 and early 1999, Crown Prince Abdullah steered his country from an attitude of cautious distancing and little contact to active engagement with Iran. In fact, improved relations with Iran have emerged as the cornerstone of Saudi Arabia's plans for regional leadership. As petroleum prices rose after the Vienna agreement, so did hopes for more breakthroughs between the two countries. These moves may benefit Iran by ending that country's isolation, but in reality the Saudis have much to gain from a more moderate Iran under a strengthened Khatami. The kingdom had a substantial, immediate short-term gain from the Iranian cooperation with the OPEC production cutback: As explained earlier, the resulting prices raised Saudi revenue by $6 billion a year, despite the lower volume of Saudi oil exports.

Saudi defense minister Prince Sultan's visit to Tehran for talks with Iranian leaders in May 1999 highlighted the distinct improvement in bilateral relations emanating from the Saudi initiative. At this meeting, Ali Shamkhani, the Iranian defense minister, told his Saudi counterpart that he saw "no limit" to ties with the kingdom and proclaimed that Iran's "entire defense capability will be put at the disposal of our Muslim Saudi brothers."[5] Later that month, when Khatami visited Riyadh, King Fahd approved the movements with an observation that "the door is wide open to develop and strengthen relations between the two countries in the interest of the two peoples and the Muslim world."[6] A state paper later noted that "it is clear that a total rapprochement in all fields, especially political and economic, is in the offing."[7] In interviews following his meetings with Khatami, the crown prince defended Iran's military buildup. "Iran has every right to develop its defense capabilities for its security without harming others. We also do the same."[8]

The United States publicly approved this relationship. Bruce Reidel, the senior director for Near Eastern and South

Asian affairs on the National Security Council, noted, "The United States is encouraged by the improvement in relations between Saudi Arabia and Iran and we welcome this important development which we consider an important and positive step towards easing tension in the region."[9] Nevertheless, many observers outside the United States believed that some in the administration had reservations about the speed with which it occurred. The only external voices of dissent came from the UAE, which has outstanding border issues with Iran, as will be discussed below.

This is not to say that the entire leadership in both countries approved of the situation; conservative religious leaders in Iran, for example, were unhappy with the rapprochement, and it seemed that they began to work behind the scenes to derail the emerging relationship. In June 1999, Iran submitted a candidate for the position of OPEC secretary general after the Saudis had nominated Suleiman Al Herbish to the post.[10] According to a Saudi ambassador, "Fielding this candidate is evidence that the senior Iranian mullahs are trying to sabotage the initiatives of the crown prince and President Khatami."[11] It is also an example of how petroleum issues are the source of both reconciliation and antagonism in Saudi international affairs.

United Arab Emirates

After Saudi Arabia, the UAE has the largest oil reserves of any Gulf Cooperation Council (GCC) member, providing the UAE with a certain share of power and until recently ensuring close relations with its neighbor, Saudi Arabia. Historically, the relationship has been very solid. It can be argued that the ruling families of Abu Dhabi and Dubai know that Saudi support has helped them to remain in power. Hence, they have generally deferred to the kingdom on matters of security and foreign policy. For example, the UAE was one of only three countries in the world (along with Saudi Arabia and Pakistan) to recognize the Taliban as the official government of Afghanistan and to fund them—and it did so at the urging of Riyadh.

Nonetheless, tensions have risen in the past few years.

First, the UAE protested the kingdom's development of Shaybah, an oil field with 14.3 billion barrels of oil and a production capacity of 500,000 barrels per day (bpd), which straddles the border between the two states. According a 1974 bilateral agreement, whichever state has more than 50 percent of the reserves on its side of the border will have complete control of the entire oil field. In the end, the UAE agreed to the Saudi position, which is well supported by geological evidence, that the overwhelming majority of the field was in Saudi territory.[12]

The challenge for Saudi Arabian oil policy is improving its cooperation on oil matters with other Persian Gulf states without exacerbating differences with the UAE. Any improvement in Saudi–Iranian relations collides directly with the dispute between the UAE and Iran over the sovereignty of the three islands of Abu Musa and the Lesser and Greater Tunbs, which has been a bilateral issue for years. According to the UAE, Iran has wrongfully occupied the islands since 1971. "Iran continues with its false and feeble claims about its right to these islands," said UAE foreign affairs minister Rashid Abdullah Al Nuaimi, "claiming that it is just a misunderstanding with the UAE. The fact is, there isn't anything called misunderstanding. This is a question of Iranian occupation."[13]

When the Iranian president visited Riyadh and the Saudis made no mention of the long-standing island dispute, a fierce war of words ensued between the Saudis and the UAE. The UAE felt betrayed and was critical of Saudi Arabia for increasing friendly communications with Iran before the issue of the islands has been settled; the Saudis believed their neighbor was overreacting. The very fate of the GCC initially appeared to be threatened by the public row that emerged between the two governments and through the state media outlets over the matter. This type of negative publicity is rare among the close-knit GCC member nations.

After the UAE publicly questioned the wisdom and fairness of the Saudi rapprochement with Iran, Saudi defense minister Prince Sultan reacted strongly: "We do not want to get into any verbal or childish disputes. Saudi Arabia's leadership and people are far above that."[14] When asked about the UAE foreign minister's comment, Prince Sultan diplomatically stated, "The Kingdom of Saudi Arabia is a fully sovereign state . . . [and] its stance does not change toward the Arab nation . . . and most especially the sisterly UAE, which is dear to us." He then added, "In any case, the ignorant is his own enemy."[15]

Within days, Qatar's emir, Shaykh Hamad bin Khalifa al-Thani, interceded to mediate a resolution, which included a commitment to "returning to the original texts when addressing Iran."[16] On June 14, 1999, the UAE and Saudi Arabia agreed to end their dispute over increasingly warm relations with Iran—in essence, the UAE conceded that it could not let the Saudi move affect Saudi-UAE relations. By the end of the month, Qatari foreign minister Shaykh Hamad bin Jassim bin Jabr al-Thani said that discussions had "destroyed any misunderstanding which might have existed and, thank God, there was total agreement."[17] A quick resolution of the disagreement was in the best interest of all involved, but some have suggested that the Saudis were encouraged in their compromise by irritation with the Iranians for nominating an OPEC secretary general candidate after the Saudis had proposed their own. A current Saudi ambassador noted, "The way the Iranians dealt with [the OPEC] secretary general issue helped us adopt a public support stance toward the Emirates on the disputed islands. . . . As they say in America, 'tit-for-tat.'"[18]

The disagreement between the UAE and Saudi Arabia illustrates a key point—that Saudi Arabia is the main ally for the GCC states and their strategic interests. In this case, Saudi Arabia did little more than hint at a unilateral movement toward Iran, and the UAE reacted very strongly. Despite its initial anger at the move, the UAE eventually had to let drop its public objections, reconfirming the importance to the UAE's long-term national security interests of strong ties with the kingdom.

Bahrain

Saudi Arabia and Bahrain have the closest relationship among the GCC nations. Bahrain has limited natural resources and few means to generate governmental income, so its well-being rests on the economic and security guarantees offered to it by Saudi Arabia. Bahrain has established itself as the banking center of the Persian Gulf, a position that Saudi investment and support allowed Bahrain to achieve. For example, Saudis have invested massive funds in major Bahrain-based banks such as the Arab Banking Corporation (the Arab world's largest bank) and Gulf United Bank, among others. The Saudis are also heavily invested in key Bahraini companies such as Balco (Bahrain Aluminum Company), Batelco (Bahrain Telecommunication Company), and several large cement factories.[19]

In addition, Saudi Arabia allocates all of the revenues from the 140,000-bpd offshore Abu Safah field to the Bahraini government.[20] This arrangement originally began with the fiction that half of the field was located in Bahraini territory, and thus the revenues were split fifty–fifty. In 1995, the Saudis transferred all of the proceeds to Bahrain—representing approximately $700 million to $800 million per year and amounting to around 45 percent of Bahrain's total 1998 budget of $1.7 billion.[21] This aid has great value for the Saudis, too. In defense and foreign affairs, Bahrain is deeply influenced by the Saudis. In fact, according to a former senior official in the Middle East department of the British Secret Intelligence Service (SIS), "Bahrain is for all intents and purposes a Saudi protectorate."[22]

An example of the Saudi influence and presence in Bahrain revealed itself in the recent Shi'i unrest there. Although the ruling Al Khalifa are Sunni, at least 70 percent of the native Bahraini population is Shi'i. Recently, when Shi'i demonstrations threatened to turn violent, as they did at other times when radicals instigated a series of fire bombings in Bahrain, the Saudi Arabian National Guard was summoned from Dhahran.[23] The Saudi troops rolled over the King Fahd causeway connecting the Saudi mainland with the island state and prepared to assist the Bahraini Security and Intelligence Service (BSIS) if the demonstration got out of hand. It ended peacefully.

The stability of Bahrain's Al Khalifa ruling family is highly important to Saudi national security. If the Al Khalifa fall, it would send a message that the Saudis do not control their immediate surroundings, fueling anti-Saudi sentiment in the Saudi Eastern Province where Shi'is constitute a slim majority—and where most of Saudi Arabia's major oil deposits are located. Any coup in Bahrain could trigger the same explosion that the 1979 Iranian revolution did, when the Shi'i population in the Eastern Province demonstrated and rioted in support of Khomeini. The appointment of the former director general of the Public Security Agency, General Abdallah bin Abdel-Rahman Al Sheikh, as the Saudi ambassador to Bahrain is emblematic of Saudi Arabia's security commitment to Bahrain's internal stability.

Iraq

Iraq is still paying the price for its attack on Kuwait, and this is stifling its development in all areas. Some analysts estimate that the resulting embargo on Iraq's oil exports through 1998 cost the Iraqis more than $130 billion in revenues.[24] The Saudis were one of the main beneficiaries of this embargo. At start of Persian Gulf War, the world lost 5 million bpd of oil production overnight, and the Saudis compensated to fill that vacuum. The kingdom increased production from 5.3 million to 8 million bpd in six months,[25] and from July 1990 to November 1996 its market share increased from 23 percent to 31 percent.[26] Meanwhile, Iraqi production hovered around 2.2 million to 2.3 million bpd, including 0.6 million bpd of smuggled crude, until late 1998.[27] Iraqi production is predicted to remain below 3 million bpd until the end of 1999.[28]

Yet, the consensus on the sanctions against Iraq seems to be eroding. Saudi Arabia must watch these developments very closely, for a return to full Iraqi production could threaten the tenuous oil price recovery created in mid-1999. In June 1999, Iraqi oil minister Amir Muhammad Rasheed announced that his country's "export capacity can even match more than 6 million bpd of oil production."[29] These comments worry other OPEC members, especially Saudi Arabia, who would lose more than others from a new oil-market flood. The possibility is even more harrowing when, as oil analyst Jamie Richard has noted, "there are scores of majors and independent companies (and banks to finance them) who are teaming to make the needed investments [in Iraq]."[30]

Surprisingly, the Saudis seem willing to allow Iraq to re-enter the market. Crown Prince Abdullah's Iraqi initiative, announced in January 1999, has two key elements: the abolition of all United Nations (UN) restrictions on the amount of oil Iraq is allowed to sell to finance humanitarian imports, such as food and medicine; and retention of sanctions, especially military sanctions, to coerce Iraq into full compliance with UN Special Committee (UNSCOM) regulations.[31] This plan indicates that the kingdom's oil policy, which can only suffer from any increase in Iraqi oil production, is not dictating the crown prince's Iraqi policy.

In key aspects, the Saudi plan overlapped with U.S. strategy and drew strong criticism from Russia, France, and other countries in the Persian Gulf region. Arab leaders held various meetings devoted to crafting a unified Arab response to Iraq, and, although Saudi diplomacy under the direction of Prince Saud Al Faisal successfully convinced Egypt of the wisdom of its plan, less powerful states such as Yemen, Qatar, the UAE, and Oman resisted, complaining about unfairness and the lack of Arab solidarity. Although by late 1999 little progress had been made on the Iraqi question, the Saudis have made it clear that they intend to be active partners in the quest for a final resolution to this problem.

Oman

Relations between Oman and Saudi Arabia have always been distant and cold. The relatively long 657-kilometer border has often been a great source of disagreement, but King Fahd and Oman's Sultan Qaboos signed a border agreement in March 1990, permitting the two nations to finalize official demarcating maps by July 1995. Even once that dispute was settled, however, religion has remained a barrier to close ties. In the eyes of the Saudis, Sultan Qaboos is not a Sunni but an

Ibadi—a movement Saudis believe is only loosely connected to Islam and will never be recognized by Saudi Arabia's puritanical Wahhabi movement.

Oman's maintenance of ties with Iraq after the Persian Gulf War and establishment of relations with Israel in 1994 further alienated the Saudis. Over the last two years, Oman has been attempting to play a greater role as mediator among the Arab gulf states on oil issues but, because of its lack of formal membership in OPEC and its small reserves, it has been only moderately successful.

Qatar

In recent years, the kingdom's relations with Qatar have been less influenced by Riyadh's oil policy than was the case in the past. In 1974, the UAE–Saudi border agreement excluded the Qataris from negotiations concerning onshore border demarcation in return for an offshore natural gas field. This issue was reopened in 1991 by Qatar, and the relationship between Qatar and Saudi Arabia has been strained ever since. In addition to this border dispute, Qatari moves such as opening communication with Iraq in mid-1992 and officially supporting the unification of Yemen during the Yemeni civil war infuriated Riyadh.[32] Although the two sides resolved many of these issues by the end of the decade, other differences appeared that have continued to challenge the relationship. Most notable among these new issues are the controversial broadcasts of the Qatari television station Al-Jazeera and the possible plan to reorganize OPEC in such a way that Qatar would lose its membership.

Al-Jazeera, founded in 1996, is a 24-hour television station that broadcasts controversial talk-shows and news to millions of viewers in the Arab world. According to Mohammad Jassem, the station's director general, "There is no subject that is off-limits to us no matter how sensitive."[33] This mission has brought the station into direct conflict with many Arab nations, who are insulted by the comments of newscasters, talk-show guests, and call-in viewers. The Saudis went on the offensive almost immediately. During the first

episode, they used their influence to quash a planned interview with Saudi dissident Mohammed Al Masari. Other shows
that dealt in an open manner with questions of women's rights
and religious issues further upset the Saudis. Finally, 1999
broadcasts of a Saddam Husayn speech and an interview with
Usama bin Ladin "infuriated the Saudi leadership," according to a former British Secret Intelligence Service station chief
in Doha. He added, "The Qataris ought to be extremely care-
ful about pushing the Saudis too far on this issue."[34]

Another Saudi–Qatari problem on the horizon is an
emerging sense that OPEC should be reorganized, with countries that produce lower volumes of oil, such as Qatar,
ultimately phased out of the group. As one Iranian oil official said, "With such small oil reserves, there really is little
justification for Qatari membership in OPEC. Replacing them
with Mexico, for example, would only strengthen the organization."[35] With only 3.7 billion barrels, Qatar is far below many
non-OPEC members in terms of proven oil reserves.

It is clear, according to Riyadh's oil policymakers, that
the Saudis will begin to pursue this reorganization aggressively in the near future.[36] The Saudi policy would advocate
incorporating as many countries as possible, including Qatar
if it wishes to remain, but with the underlying overt notion
that decisions will be made strictly by a core group of large
producers. This is likely to become another thorn in the side
of relations between the two Gulf states, and is therefore an
issue that will be revisited in the conclusion chapter.

Kuwait

Kuwait and Saudi Arabia have long had a strong relationship,
founded on Kuwait's security needs. The Iraqi invasion of Kuwait and vital Saudi support in that conflict have brought the
two countries even closer together. For example, in OPEC
meetings, Kuwaitis almost always follow the Saudi lead, saving
their criticism of Saudi oil policy for private surroundings. According to a former Kuwaiti ambassador, "We clearly realize
that the Saudis, along with the United States, are our closest—
and really, when push comes to shove, our only two—allies."[37]

A concrete manifestation of this partnership is the joint sovereignty of the "Divided Zone" on the border between the two countries. This oil-rich zone containing 5 billion barrels of proven reserves is controlled by Saudi Arabia and Kuwait and produced by three companies—Saudi Arabian Texaco, Kuwait Petroleum Corporation (KPC), and the Arabian Oil Company (AOC). The Divided Zone concession that Saudi Arabia Texaco operates is the onshore, Saudi portion of the zone; KPC operates the onshore section in Kuwait; and the AOC operates offshore for both sides. Saudi Arabia and Kuwait formalized their respective rights to this zone in July 1965 in an agreement based on the Uqair Conference of 1922. At this conference, the British, negotiating on behalf of their protectorates Kuwait and Iraq, and Ibn Saud demarcated the border between the two countries. The British officially recognized Saudi control of the entire Eastern Province in return for Saudi concessions on the Iraqi border and official recognition of Kuwaiti sovereignty.[38]

Relations between Saudi Arabia and Kuwait have been cemented by common concern about the threats facing them, especially from Iraq—evidenced by the Iraqi plans for aggression in 1967, which were halted only through swift British intervention; the Iran–Iraq war of 1980–88; and the 1990 invasion, when the ruling Al Sabah family sought refuge in Saudi Arabia. Since then, the two countries have had a "Pact of Steel" in dealing with the Iraqi regime. They have been the most active and outspoken critics of Saddam Husayn in the Middle East. Although ordinary Kuwaitis may resent their "little brother" status relative to Saudi Arabia, Kuwaiti rulers know that they must maintain extremely close relations with Saudi Arabia to deter the existential threat to Kuwait from Iraq.

Organization of Petroleum Exporting Countries

Saudi Arabia was instrumental in the creation of OPEC, which was founded in 1960 to establish a strategy countering the unilateral decision of the major oil companies to lower "posted prices" of Arab Light from $1.90 to $1.76 per barrel.[39] Its record of effectiveness since then has been spotty. Now, at the end of the 1990s, it has become less relevant than ever because Saudi Arabia and a few key OPEC and non-OPEC producers are slowly growing more capable of managing prices on their own. Potential quota-busters are kept at bay by the kingdom's unprecedented 3.2 million bpd of excess production capacity, which, if unleashed, would drive down prices and drive many OPEC producers out of business. For this reason, some oil analysts have noted that "OPEC is becoming Saudi OPEC."[40] Such aggressiveness was seen at Jakarta in November 1997, when the Saudis single-handedly pressured fellow members to increase production, and most recently at the Vienna meeting in March 1999, when they convinced every member to cut production.

The Saudis constructed the Vienna deal before the conference and without substantial input from any other OPEC member except Iran and to some extent Venezuela, providing one more piece of evidence that the cartel may be drifting toward irrelevance. No one denies that the Vienna meeting merely rubber-stamped what the Saudis and Iranians had achieved separately or that the Saudis once again wield unquestionable power in the cartel. As one senior Nigerian oil official said, "You Saudis are the only ones who could one day drive us all [the Venezuelans, Mexicans, Norwegians and all non-Middle Eastern countries] out of business."[41] Another sign of this growing power was the announcement by Oil Minister Ali Al Naimi in early 1999 of the formation of an "informal" group of oil producers—the main exporters, regardless of OPEC membership, and hence a not-so-subtle threat to the current composition of OPEC.

Although this informal, *de facto* control has always existed, Saudi Arabia has been more assertive in gaining *de jure* control over OPEC in recent years. It has always operated with a strong hand behind closed doors but is now moving more openly to solidify its position as the preeminent OPEC member. After the June 1999 nomination of Sulaiman Al Herbish to the post of OPEC secretary general, a contest quickly appeared between the kingdom and Iran, which nominated Hosein Kazempour Ardibili, Iran's current OPEC governor, for the position. As neiter side withdrew its candidate and the position requires a unanimous vote of the members, the selection remained undecided after the subsequent OPEC meeting. Neither producer has held the secretary general spot in more than thirty years.

Venezuela and Mexico

The Saudis have long been suspicious of the Venezuelans and have accused them of cheating on production agreements to increase their revenues and solidify their position in the U.S. market. Yet, the drastic price drop of 1998 forced the Saudis to overlook these grudges and try to find common ground. In March 1998 Prince Abdulaziz bin Salman, the deputy oil minister for petroleum affairs, was approached by Adrian Lajous, the chief executive of Pemex, Mexico's national oil company, who suggested a meeting of Venezuelan, Saudi, and Mexican officials. Prince Abdulaziz approved the idea of having Mexico, a non-OPEC member, mediate between the other two, which are members of the cartel. The prince thus orchestrated one meeting in Riyadh and one in Amsterdam. According to Mexican energy secretary Luis Tellez, Mexico volunteered to "act as a middleman" between the Saudis and the Venezuelans.[42] The result was a production reduction agreement among these three main exporters to the U.S. market. Although the two meetings were unable to stem the drop in prices, they laid the groundwork for the ultimately successful Hague meeting in March 1999.

Other positive results of these negotiations emerged quickly. In a speech at Johns Hopkins University's School of Advanced International Studies on March 1, 1999, the new Venezuelan oil minister, Ali Rodriguez, announced that his country would not attempt to recapture the market position it lost to Saudi Arabia in 1998. "We're not going to be in competition with Saudi Arabia," said Rodriguez.[43] The Mexicans, too, seemed committed to working under the Saudi-led production reduction regime. In a joint statement released in Riyadh, Al Naimi and Tellez "reaffirmed their commitment of cooperation with other producers to take whatever action is needed to stabilize the market and improve the oil prices."[44]

Yet, difficulties remained. By mid-1999, Venezuelan president Hugo Chavez began speaking of plans to create a Latin American OPEC composed of Venezuela, Mexico, and Brazil. Venezuela's OPEC governor revealed the country's thinking when he stated that "if the 'old' OPEC, dominated by Saudi Arabia (which, since the end of the Gulf War, has become subordinate to the United States), is to keep on playing a significant role . . . the creation of a new organization will be necessary."[45] The solution he offered was a new Latin American organization: "In this way, we can have a more balanced say on the price of oil, by taking a unified stance and speaking in one voice for all Latin American producers, which represent nearly 15 percent of the total world production of crude."[46] Needless to say, the Saudis were not pleased with such talk.

In this case, the demands of Saudi oil policy—that is, the need to strengthen prices—led to rapprochement between former disputants. Moreover, given their positions as the three main exporters to the United States, the three countries' relations can be expected to remain a priority in the future.

Notes

1.   Anthony Cave Brown, *Oil, God, and Gold: The Story of Aramco and Saudi Kings* (Boston: Houghton Mifflin, 1999), p. 344.

2.   "Iran's Khatami Seen Strengthening Saudi Ties," Reuters, May 19, 1999.

3.   "Saudi Airlines to Resume Flights to Iran," Reuters, February 22, 1999.

4.   See Michael Georgy, "Saudi Desert Offers Oil Investment Guide," Reuters, February 22, 1999; "Crude Oil Steady as Traders Watch for More OPEC Cuts," Bloomberg, February 22, 1999.

5.   "Iran's Shamkhani: No Limits to Ties with Saudi Arabia," *Tehran IRNA*, May 1, 1999 (translated by Foreign Broadcast Information Service).

6.   Agence France-Presse, "Fahd Foresees Good Outlook for Iran Ties," *International Herald Tribune*, May 17, 1999, p. 2.

7.   Ibid.

8.   "Saudi Prince Says Iran Has Right to Arm Itself," Reuters, June 1, 1999.

9.   "U.S. Encouraged by Saudi-Iran Rapprochement—Paper," Reuters, June 10, 1999.

10. "Iran to Submit Candidate to Rival Saudi for OPEC's Top Job," Reuters, June 27, 1999.

11. Author interview by telephone, July 2, 1999, in Washington.

12. Olivier Da Lage, *Geopolitique de L'Arabie Saoudite* [Geopolitics of Saudi Arabia] (Paris: Editions Complexe, 1996), pp. 122–123.

13. "Oman Said Narrowing Saudi–UAE Differences over Iran," *BBC Summary of World Broadcasts,* June 15, 1999 (translation from article on Al-Hayat website, June 13, 1999).

14. "Tensions Resurface between Saudi Arabia and UAE," *Mideast Mirror* 13, no. 106 (June 7, 1999).

15. Ibid.

16. "Gulf Arabs Support UAE in Islands Row with Iran," Reuters, July 3, 1999.

17. "Qatar Says UAE-Saudi Row over Iran Resolved," Reuters, June 19, 1999.

18. Author interview in Washington, June 29, 1999.

19. Da Lage, *Geopolitique de L'Arabie Saoudite,* pp. 124–125.

20. Energy Intelligence Group (EIG), "Saudi Aramco," part of EIG's series *The World's Key National Oil Companies,* January 1999, p. 24.

21. CIA Factbook, located online at: http://www.odci.gov/cia/publications/factbook/ba.html.

22. Author interview in London, February 2, 1999.

23. Ibid.

24. Raad Alkadiri, "Iraq Under Sanctions: Diminishing Returns," in Rosemary Hollis, ed., *Oil and Regional Developments in the Gulf* (London: Royal Institute of International Affairs, 1998), p. 93.

25. EIG, "Saudi Aramco," p. 7.

26. *Energy Compass* 9, no. 46, p. 2.

27. Fadhil Chalabi, "Iraq's Future Oil Supplies and Their Impact on Oil Prices," *World Oil Prices* (London: Center for Global Energy Studies, 1998), p. 130.

28. "Iran OPEC Governor Says Too Soon to Review Cuts," Reuters, June 2, 1999.

29. Hassan Hafidh, "Iraq Says Its Export Facilities Have 6 Mbpd Capacity," Reuters, June 24, 1999.

30. Barry Rubin, ed., "An Exchange on Oil Prices," *Middle East Review of International Affairs* (*MERIA*), no 6. June 1999.

31. Patrick Clawson and Nawaf Obaid, "Assessing Proposals for Changing UN Restrictions on Iraq," *PolicyWatch,* no. 362 (The Washington Institute for Near East Policy), January 19, 1999.

32. Da Lage, *Geopolitique de L'Arabie Saoudite,* pp. 123–124.

33. Diana Abdallah, "Daring Qatari TV Station Jolts Arab World," Reuters, May 21, 1999.

34. Author interview in London, April 15, 1999.

35. Author interview in Vienna, March 24, 1999.

36. Author interview in Riyadh, October 17, 1999.

37. Author interview in Geneva, April 5, 1999.

38. Da Lage, *Geopolitique de L'Arabie Saoudite,* pp. 122–123.

39. David G. Heard, "Development of Oil in the Gulf: The UAE in Focus," in Rosemary Hollis, ed., *Oil and Regional Developments in the Gulf* (London: The Royal Institute of International Affairs, 1998), p. 45.

40. Michael Georgy, "Saudis Seek Even Stronger OPEC Position," Reuters, June 23, 1999.

41. Author interview in Geneva, March 20, 1999.

42. "Mexico Sees $18–$20 Target for Brent Crude Oil," Reuters, June 11, 1999.

43. "Venezuela Concedes Defeat in Battle for U.S. Oil Market Share," *Stratfor,* March 4, 1999, located online at www.stratfor.com.

44. Fahd Al-Frayyan, "Saudi, Mexico Optimistic on Oil Cut Deal," Reuters, June 5, 1999.

45. *Le Point,* May 21, 1999, p. 32.

46. Ibid.

Chapter 8

Relations with Oil Consumers

S audi Arabia has worked recently to improve its relations with the world's main oil-consuming nations. It has, for example, vigorously pushed for contracts and sales in Asia, which the kingdom believes will be the fastest-growing market in the coming decade. It has also regained its position as the number-one supplier to the United States, a position it had lost for several years to Venezuela. That the kingdom has worked so hard to maintain strong relations with consumers is more evidence that it is not prepared to embark on a "flood-the-market" strategy—that is, increasing its production to the point that oil prices fall so low that many non-Saudi producers are driven out of business. If it were to follow a flood-the-market strategy, Saudi Arabia could rely on low prices alone to guarantee its access to markets in the consuming countries. But because the kingdom seems intent on maintaining prices at the $18–$20 level, it can expect to have a great deal of competition. Thus, proactive marketing and sales arrangements, as well as investments in petroleum projects with consuming nations, will take on greater importance.

The United States

Ever since King Abdulaziz and President Franklin Roosevelt met on the U.S.S. *Quincy* in 1945, the United States and Saudi Arabia have enjoyed what both countries call a "special relationship." This strategic alliance, which is one of America's most enduring and one of the closest the United States has with any Middle Eastern country, has never been severed and only rarely strained. Built on a long tradition of close economic and military ties, the foundation of this relationship is the kingdom's possession of the world's largest supply of oil reserves (at least

260 billion barrels in 1999) and the continuing U.S. reliance on foreign crude imports. To the end of the century, the kingdom remained a "strategic vital ally," according to David Welch, assistant secretary of state for international organization affairs and a former acting chief of mission in Riyadh.[1] Although the United States and Saudi Arabia do not share a formal defense treaty, nearly every administration has made it clear that maintaining strong ties to the kingdom is one of Washington's top foreign policy priorities.

One of the most important milestones in U.S.–Saudi relations was the oil embargo of 1974, which had long-term negative consequences for the United States and much of the industrialized world. At first glance, it may seem that the embargo reflects the tenuousness of the partnership between

Key U.S. Statements

"An attempt by any outside force to gain control of the Persian Gulf region will be regarded as an assault on the vital interests of the United States of America and such an assault will be repelled by any means necessary, including military force."

—President Jimmy Carter, January 23, 1980
(The Carter Doctrine)

"There's no way that we could stand by [and see Saudi Arabia] taken over by anyone that would shut off that oil."
—President Ronald Reagan, October 1, 1981
(The Reagan Corollary)

"The sovereign independence of Saudi Arabia is of vital interest to the United States."
—President George Bush, August 8, 1990
(following Iraq's invasion of Kuwait)

"President Clinton's commitment to the security of friends in the Gulf, like that of every president since Franklin Roosevelt, is firm and constant."
—Secretary of State Warren Christopher
(February 21, 1993)

the United States and the Saudis. Upon closer examination, however, it becomes clear that King Faisal resorted to the "oil weapon" only after enormous domestic pressure (mainly from the religious establishment) and even then took steps to mitigate the negative impact he knew this policy would have on the United States.[2]

The origins of the embargo lay with the June 1967 Arab–Israeli War, when Israel won a devastating victory over the armies of Egypt, Jordan, and Syria and also occupied substantial territory. In the following years, the Arabs searched for ways to pressure Israel into returning this land. Many proposed an oil embargo on countries supporting Israel. At first, King Faisal dismissed the prospect of a boycott, saying it was useless to talk about the use of oil as an instrument of pressure against the United States; indeed, it was dangerous even to think along those lines, because the Saudis did not know how Washington would react.[3] By April 1973, however, the king began sending signals to the U.S. government that, unless some progress was made regarding Israel's withdrawal from the lands it took in 1967, the Arabs would use their oil for political leverage.[4]

In 1973, the Egyptians attacked Israeli positions along the Suez Canal in an attempt to regain land lost in the 1967 war. This attack had King Faisal's approval, and he agreed to use the only effective weapon he had—oil—if the war went badly.[5] The Saudi monarch realized that an embargo might hurt Saudi interests, given his country's need for U.S. security support. He might have also worried about the long-term implications of the rise in oil prices that would accompany an embargo.

King Faisal had ample reason to be cautious in his use of oil as a weapon and to delay its use as long as possible. He could afford to ignore domestic pressure while the Egyptians at least seemed not to be losing the war. On October 19, 1973, however, news that the Egyptians faced a near-total defeat reached the Saudi king and forced his hand; the next day, he announced the embargo. The official announcement read: "In view of the increase in support for Israel, the Saudi Arabian Kingdom has decided to stop the export of oil to the

United States of America for adopting such a stand."[6] The embargo lasted until the early 1974 disengagement agreement was signed.

Yet, even during the embargo, large technical, military, and economic cooperation programs between the United States and the kingdom continued unaffected, revealing that the two sides were committed to limit the fallout and to ensure that political differences did not threaten to derail the long-term bilateral relationship. Furthermore, despite publicly urging a total ban on oil sales to the West, the Saudis continued clandestinely to ship oil to the U.S. fleet in the Mediterranean during these crucial months.[7] They did so at a considerable political risk—if nothing else, they could easily have been accused of hypocrisy. In some ways, the embargo highlights the resilience of, as well as the pitfalls in, the U.S.–Saudi partnership.

As the new century begins, the twin pillars of this "special relationship" are still trade and security. Saudi Arabia is by far America's largest trading partner in the Middle East. The United States, which consumes 25 percent of the world's av-

U.S. Oil Imports From Major Crude Suppliers				
COUNTRY	**2/99**	**3/99**	**4/99**	**Avg. '98**
Saudi Arabia	1.437	1.584	1.398	1.386
Venezuela	1.291	0.998	1.356	1.356
Mexico	1.231	1.426	1.313	1.304
Canada	1.082	1.053	0.970	1.112
Nigeria	0.661	0.630	0.865	0.679
Iraq	0.681	0.791	0.840	0.334
Colombia	0.458	0.572	0.425	0.326
Angola	0.333	0.283	0.409	0.446
Kuwait	0.205	0.324	0.279	0.279
Norway	0.157	0.200	0.191	0.216
TOTAL	**7.536**	**7.861**	**9.054**	**8.547**

Note: All quantities in millions of barrels per day
Source: "Saudi Leads as U.S. Oil Imports in April Run Ahead of 1998," Reuters, June 14, 1999.

erage daily output of oil, imports more than half its oil, and Saudi Arabia is the largest single supplier. Saudi exports to the United States totaled $9.6 billion in 1997, compared with $8.8 billion in 1996. By mid-1999, the United States was importing more crude oil from Gulf Cooperation Council (GCC) states than at any time since recordkeeping began in 1972.[8]

At the same time, the United States is the largest of Saudi Arabia's trading partners, accounting for 22 percent of all Saudi imports.[9] According to the U.S. Census Bureau, Saudi Arabia's total imports (excluding military transfers) from the United States in 1998 increased to $10.5 billion from $8.4 billion in 1997.[10] Military transfers account for even more Saudi imports. From 1985 to 1996, the kingdom imported more than $30 billion worth of U.S. military hardware.[11]

Saudi Arabia's commitment to the U.S. market is strong.

U.S. Arms Transfers to Saudi Arabia, 1950–1997*				
Category	**Orders** $ billions**	**% of total orders**	**Deliveries $ billions**	**% of total deliveries**
Weapons and Ammunition	19.893	21.2	9.092	15.6
Support Equipment	16.614	17.7	9.815	16.8
Spare Parts, Modifications	9.778	10.4	5.259	9.0
Supply, Repair, Training	29.615	31.6	17.804	30.6
Construction	17.924	19.1	16.197	27.8
TOTALS	**93.824**	**100.0**	**58.167**	**99.8**

* Figures are current through March 31, 1997; source: Alfred B. Prados, "Saudi Arabia: Post-War Issues and U.S. Relations" (Washington: Congressional Research Service, Library of Congress, November 13, 1998).
** Orders are larger than deliveries because many of the weapons systems ordered have not yet been delivered.

Its first joint venture was Star Enterprises, established with Texaco in 1988. Ten years later, Saudi Aramco established Motiva Enterprises LLC with Texaco and Shell, with stakes of 32.5 percent, 32.5 percent, and 35 percent, respectively. Motiva has assets of $13 billion dollars[12] and controls thirteen refineries, of which the four largest have a combined capacity of 820,000 barrels per day (bpd).[13] It also operates forty-eight bulk plants and loading terminals and 22,000 service stations that sell 600,000 bpd of gasoline.[14] Motiva accounts for 15 percent of U.S. refining capacity and 20 percent of the U.S. gasoline market.[15] This investment, the largest downstream venture in U.S. refining and marketing networks, guarantees the kingdom greater access to the market of the world's largest oil consuming country. "From an international standpoint, this is an extremely significant development for Saudi Aramco," said Saad Al Shaifan, Aramco senior vice president for international operations, "because it positions us as a full partner with two world class oil companies in a well-established market."[16] One U.S. oil company official said confidentially that his colleagues believe the Saudis give Motiva a slight discount on crude, showing the significance they place on the U.S. market.[17]

The extent of the two countries' economic interrelations seems poised to grow. With the invitation for U.S. oil companies to reenter the kingdom's energy sector and the general liberalization of the Saudi economy to outside investment, many opportunities for U.S. exports have appeared. As mentioned in the previous chapter, the Saudis in 1999 recaptured from Venezuela the position of top U.S. oil supplier. David Welch, assistant secretary of state for international organization affairs and a former acting chief of mission in Riyadh, has likewise pointed out, "Regardless of market share fluctuations, over the strategic horizon, Saudi Arabia will remain the most important oil supplier to America."[18]

This economic relationship has strong security components as well. For more than fifty years, the two governments have found it mutually beneficial to maintain strong security bonds. Although a complete examination of these ties are beyond the scope of this book, the billions of dollars in yearly purchases allowed Saudi Arabia to remain the world's largest single importer of U.S. military equipment throughout the 1990s. In addition, approximately 5,000 U.S. troops remain stationed in the kingdom.[19]

Despite these bonds, there are challenges to the partnership. In mid-1999, an independent group of small- and medium-sized U.S. producers filed an antidumping petition that accused Saudi Arabia, Mexico, Venezuela, and Iraq of selling oil to the U.S. market at prices lower than their domestic market prices and that sought to have import duties placed on these countries' oil exports to the United States, which together total between 4 million and 5 million bpd.[20]

The strongest disagreements between Washington and Riyadh, however, are reserved for issues relating to Israel. When the United States first recognized the state of Israel in 1948, King Abdulaziz declared that the Americans are "no longer our brothers but our business partners."[21] More than half a century later, in June 1999, Crown Prince Abdullah spoke with equal passion about the remote possibility for peace with Israel: "We will not accept part or half solutions even if the whole world signed them. Either our full rights, foremost of which our holy Jerusalem, or peace will remain in the womb of the unknown."[22] Despite the strength of its feeling on this issue, Saudi Arabia took part in the peace process begun in Madrid in October 1991, participating in numerous sessions at the multilateral peace talks. After the government of Ehud Barak took office in Israel in 1999, Saudi diplomats also took part in a U.S.-arranged session with their Israeli counterparts at the United Nations.

Finally, although the State Department has quietly approved the warming of relations between Saudi Arabia and Iran, some officials admit that they are "not very happy with the kingdom's unilateral approach to rapprochement."[23] As long as the United States views Iran as a rogue state that supports terrorism, it is hard to imagine that the United States will completely support closer ties between Iran and the kingdom.

East Asia

Before the Asian economic crisis of late 1997, Asia was Saudi Arabia's largest crude export market, totaling 60 percent of sales.[24] Although Japan and other Asian nations suffered depressed economies in 1998, by mid-1999 there was hope that the worst was over, and indeed the region did resume economic growth in subsequent months. Studies conducted in the middle of the crisis continued to estimate that East Asian demand for oil would increase by 42 percent from 1995 to 2010.[25] According to Aramco chief executive Abdallah Jumah, "The Far East continued to be the company's largest energy market and the region of greatest prospective growth, despite turmoil in financial markets in the region during the last half of 1997."[26]

The Saudis have already capitalized on this trend by establishing joint ventures with Asian companies. In 1991, Saudi Aramco entered the South Korean market with the purchase of a 35 percent stake in two of Ssang Yong's oil refineries, which each have the potential to refine up to 300,000 bpd.[27] In 1994, Aramco bought 40 percent of Petron Corporation, the largest refiner in the Philippines.[28] In Japan, too, the Saudis have a large interest, because the Japanese are currently the largest importer of oil in the East Asia region,[29] and the majority of their petroleum comes from the Middle East— 85.9 percent in May 1999, of which 21 percent came from Saudi Arabia.[30] For their part, the Japanese are 80-percent stakeholders in the Arabian Oil Company, holders of the concession to the offshore reserves of the Divided Zone between Saudi Arabia and Kuwait.[31] It remains to be seen whether the Japanese will regain this concession after it expires in 2000 and 2003 on the Saudi and Kuwaiti sides, respectively, but in any event Tokyo and Riyadh will continue to be important trading partners.

The Saudis realize, however, that in the long term, China is likely to become their leading market. China's energy needs are growing faster than any other country in the region. Between 1994 and 1997, China's reliance on Persian Gulf oil rose from 40 percent to 60 percent of its total imports; its oil imports from the Gulf may surpass 90 percent by the year 2010, of which at least one-third will come from Saudi Arabia.[32] In 1997 alone, China's oil consumption rose 9.4 percent.[33] For this reason, Saudi leaders have focused much of their efforts on establishing themselves in Chinese markets, and the Chinese similarly have focused on creating partnerships with the Saudis. In 1998, Saudi Aramco conducted a joint feasibility study for the kingdom's third downstream venture in the Far East. This study explored the possibility of working with affiliates of Exxon and China Petrochemical Corporation (SINOPEC) to develop a $3 billion integrated petroleum and petrochemical facility at the existing petrochemical refinery in Fujian province.[34] A strengthening of the relationship between the world's largest potential importer and the world's largest exporter should be expected.

European Union

Over the past decade, the Saudis have made several key purchases in European oil companies. One example is the purchase of two 265,000-bpd-capacity refineries from the Swedish firm OK Petroleum. A Saudi businessman of Yemeni origin who represents certain members of the Saudi royal family and is loosely affiliated with Nimr Petroleum, bought the refineries in 1994 for $1.2 billion.[35] In 1996, Saudi Aramco finalized the purchase of 50 percent of Greece's Motor-Oil Hellas Refining and Marketing Petroleum Company from the Vardinoyannis shipping family. With this deal, valued at approximately $650 million, Saudi Aramco also acquired Motor-Oil Hellas' subsidiary, Avinoil, which gave the company an extremely strong presence in Eastern Europe, especially the Balkans.[36] Also worth noting is the Al Yamamah deal, finalized in 1985 between Saudi Arabia and Great Britain (Al Yamamah II was signed in 1993); this agreement is the world's largest oil-for-arms deal, valued at £30 billion to £35 billion ($47 billion to $55 billion). Under the 1996 contract, the Saudis received military hardware such as Tornado fighters and training services in exchange for oil deliveries of 600,000 bpd.[37] By 1999, in response to the scaling back of the oil-for-arms deal and also to keep Saudi exports in line with the OPEC production cuts, the deliveries had dropped to 300,000 bpd.[38]

As the Saudis continue to focus on the U.S. and Asian markets, the European Union is becoming less important to them as an export destination. High taxes, closer sources of petroleum for European consumers (such as Libya and the North Sea), and strong ties to other oil exporters (such as Iraq and Algeria) ensure that Europe will continue to decline in importance as an importer of Saudi crude oil. The strategic importance of these countries to Saudi Arabia will likely diminish as well.

Currently, the British rely on the North Sea for their oil, and Norway is the prime supplier for the continent. The potential for Libya to resume petroleum production in the near future makes it even more likely that Europe will turn to this North African country to supply its energy needs. Libya's main partner in this trade will be the Italians because of proximity and historical, colonial ties. Within three to four years, the Libyans may be able to achieve their earlier levels of production capacity, and once this occurs, the Saudis will have even less of a chance of achieving a major European market share. Finally, France's Total and ELF Aquitaine will dominate in Iraq and Algeria, and cooperative agreements between these countries can be expected. The French have been aggressively pursuing the Iraqi market despite UN sanctions. High tariffs are another barrier to Saudi Arabia increasing its market share in Europe; the Italian government is currently earning more per barrel than the Saudis are, because of their relatively high petroleum tax rates.[39]

Notes

1. David Welch, U.S. assistant secretary of state for international organization affairs, interview with author in Washington, May 26, 1999.

2. For a more complete examination of this topic, see Nawaf Obaid, "The Power of Saudi Arabia's Islamic Leaders," *Middle East Quarterly* (September 1999), p. 51.

3. David Golub, *When Oil and Politics Mix: Saudi Oil Policy, 1973–1985* (Cambridge, Mass.: Center for Middle Eastern Studies, Harvard University, 1985), p. 8.

4. David E. Long, *The Kingdom of Saudi Arabia* (Gainesville: University Press of Florida, 1997), p. 69.

5. Obaid, "The Power of Saudi Arabia's Islamic Leaders," p. 52.

6. *New York Times*, October 21, 1973, p. 28.

7. Anthony Cave Brown, *Oil, God, and Gold: The Story of Aramco and Saudi Kings* (Boston: Houghton Mifflin, 1999), p. 297.

8. Tom Doggett, "U.S. Buying Persian Gulf Oil at Record Pace in 1999," Reuters, June 23, 1999.

9. See the Central Department of Statistics, *Directory of National Statistics, 1986–1990* (Riyadh: Ministry of Finance and National Economy, 1986–1990); U.S. Census Bureau, 1991–1997.

10. Ibid.

11. Directorate for Information Operation and Reports, Foreign Military Sales Program, *Foreign Military Sales, Foreign Military Construction Sales and Military Assistance Facts, 1996* (Washington: U.S. Department of Defense, 1996).

12. Edmund O. Sullivan, "The Kingdom Faces the Future," *Middle East Economic Digest*, January 22, 1999, p. 33.

13. Ibid.

14. *Saudi Aramco Dimensions Quarterly* (Fall/Winter 1998), p. 27.

15. Sullivan, "The Kingdom Faces the Future."

16. *Saudi Aramco Dimensions Quarterly* (Fall/Winter 1998), p. 27.

17. Author interview in San Francisco, August 10, 1999.

18. Author interview in Washington, May 14, 1999.

19. John Diamond, "U.S. Approves Missile Sale to Saudis," Associated Press Online, March 7, 1999.

20. William Maclean, "Oil Soars as Bulls Raise the Tempo," Reuters, July 1, 1999.

21. Brown, *Oil, God, and Gold*, p. 203.

22. "Saudi Asks Israel's Barak to Show Good Intentions," Reuters, June 1, 1999.

23. Author interviews in Washington, spring 1999.

24. Anthony Cordesman, "Economics, Energy and the Future Stability of Saudi Arabia," *Strategic Energy Initiative*, Center for Strategic and International Studies, February 10, 1999.

25. Koko Kanno and Keiichi Yokobori, "Oil and Gas Demand in Asia Pacific to 2010," in *Oil and Gas Investments in Asia under Conditions of Oil Price Uncertainty and Financial Crisis* (London: Centre for Global Energy Studies, 1998), p. 75.

26. *Saudi Aramco Dimensions Quarterly* (Winter 1997–1998), p. 16.

27. See International Energy Agency (IEA), *Middle East Oil and Gas* (Paris: Organization of Economic Cooperation and Development/IEA, 1995).

28. *Saudi Aramco Dimensions Quarterly* (Fall/Winter 1998), p. 30.

29. Kanno and Yokobori, "Oil and Gas Demand in Asia Pacific to 2010," p. 80.

30. "May Petroleum Product Sales Rise 0.6 Percent, Up 4th Month," Kyodo News Service, June 25, 1999.

31. Energy Intelligence Group (EIG), "Saudi Aramco," part of EIG's series *The World's Key National Oil Companies,* January 1999, p. 27.

32. Kang Wu and Fereidun Fesharaki, cited by *Oil and Gas Journal* 95, no. 35 (August 28, 1995); author's estimates.

33. Fereidun Fesharaki, "Developments in the Chinese Oil Industry and the Reform Programme," in *Oil and Gas Investments in Asia under Conditions of Oil Price Uncertainty and Financial Crisis* (London: Centre for Global Energy Studies, 1998), p. 201.

34. *Saudi Aramco Dimensions Quarterly* (Winter 1997–1998), p. 16.

35. See International Energy Agency (IEA), *Middle East Oil and Gas* (Paris: Organization of Economic Cooperation and Development/IEA, 1995).

36. EIG, "Saudi Aramco," p. 33.

37. Patrick Fitzgerald, "The Scandal of al-Yamamah," *New Statesman and Society,* January 12, 1996.

38. *Financial Times Middle East Energy Newsletter,* March 1999, p. 3.

39. See IEA tax tables for 1998–99.

Chapter 9

Conclusions

An examination of the Saudi oil industry, from its main actors to its current effect on the kingdom's foreign policy, reveals several insights. First, and most important, the crown prince and senior Saudi leadership have adopted an increasingly professional approach to policymaking and implementation, an approach that has influenced key ministries (especially the oil ministry). They have also placed greater faith in the private sector and summoned a growing political will to help it expand. Finally, the senior Saudi leaders are prepared to take a more assertive stand in regional and international politics.

Abroad, these policies appear to be bearing fruit with countries as varied as Iran, Libya, and the United States. Domestically, the senior leadership has avoided the worst effects of a potentially debilitating economic crisis. Much credit for these successes belongs to the senior Saudi leadership and its teams of technocrats. With the new policies they have established, however, will come new problems. Most important, Saudi assertiveness in international affairs may put pressure on its relations with the United States and the Organization of Petroleum Exporting Countries (OPEC). The recent decline in oil prices and resultant economic recession emphasized that the time is ripe for Saudi Arabia to rethink its oil policy.

Ensure the Continuation of the U.S.–Saudi 'Special Relationship'

Despite drastic differences in religion, culture, and politics, the United States and Saudi Arabia have found a way, inspired by mutual need and shared strategic goals, to coexist in respectful friendship. Even in times of great stress, the allies have always shown the strength of their "special relationship." Saudi Arabia's growing assertiveness in Persian Gulf and Arab affairs will, in the end, lend the region stability and growth. Yet, some members of the U.S. policymaking establishment will perhaps perceive Saudi Arabia's emerging assertiveness and independence in foreign affairs as a threat to the partnership. To the extent that Saudi Arabia develops an assertive, professional government, it may be less willing than in the past to accept U.S. advice automatically. Although that development has its disadvantages for the United States, the advantages derive from a Saudi government that better reflects the views of Saudi public opinion. If policymakers in both states realize that it is natural for allies to disagree on some issues secondary to their relationship, in the long run, their interests will be best served by supporting one another.

If the Saudi monarchy were seriously undermined, the delicate power sharing between the royal family and the religious establishment (closely beholden to the extreme wing, led by Shaykh Salman Al Awda, Shaykh Safar Al Hawali, and other fundamentalist clerics), would be unbalanced. So-called "charities," such as the World Muslim League, International Islamic Relief Organization, Islamic Development Bank, Organization of the Islamic Conference, Saudi Red Crescent, and dozens of other organizations, would veer toward fundamentalism and extremism without the moderating influence of the royal family, in turn opening a veritable Pandora's box of troubles on which the current Saudi regime keeps a tight lid. The most effective way Washington can counter this threat is by supporting close ties between Saudi Arabia and its fellow Arab and Muslim countries and by ensuring that the special partnership Washington and Riyadh have cultivated over decades is not jeopardized.

In the past two years, Crown Prince Abdullah—with the support of Minister of Defense Prince Sultan, Interior Minister Prince Nayef, Governor of Riyadh Prince Salman, and Vice Interior Minister Prince Ahmad—has shown a much-needed, increased pragmatism and fiscal conservatism. U.S. encouragement of more foreign direct investments and more joint ventures in offset programs (such as those in the downstream petrochemical industry) will consolidate the efficacy of the crown prince's actions.

At the same time, U.S. policymakers should realize that consensus may not be achievable on all issues—especially those involving Israel; the sanctions on Iraq; and U.S. policy toward Syria, Iran, and Pakistan. Nevertheless, they should not allow potential differences to obscure the benefits to the United States of maintaining ties to one of its most important allies in the developing world.

Reorganize OPEC

Many observers have noted the irrational nature of OPEC membership. At its founding in 1960, OPEC contained the key large petroleum exporters—Saudi Arabia, Iraq, Iran, Kuwait, and Venezuela. But today, relatively small exporters such as Algeria, Qatar, Nigeria, and Indonesia are also members, while much larger exporters—Russia, Mexico, and Norway—are not, making global production management difficult and creating an irrational distribution of power within what is meant to be a cartel of the world's oil exporters. As a Nigerian oil official said at the March 1999 Vienna OPEC conference: "Every time I walk past the Mexicans or the Norwegians, on my way to an OPEC meeting, I feel a little guilty."[1] If Saudi strength in the world oil industry manifests itself in a desire to reform this organization, such a move would benefit all exporters, regardless of whether they formally took part in any new supply-management group.

As the United States is the world's largest oil consumer and importer, U.S. interests would be hurt by sustained higher oil prices, were the cartel able to achieve that. Nevertheless, it could be argued that the United States should be prepared to accept slightly higher oil prices—which may be all that the cartel could achieve—in return for the benefits to the United States of greater price stability. This would be partially useful to the high-cost U.S. oil industry, which is hurt badly by price savings. Also, any extra oil income would accrue largely to

U.S. allies, such as Mexico and Saudi Arabia, and to other countries such as Russia, whose stability is an important U.S. interest.

Aside from having low export levels, some OPEC countries have much smaller proven petroleum reserves than other member states. For example, Qatar's reserves are 3.7 billion barrels, whereas Mexico, which is not an OPEC member, has reserves of 40 billion barrels. To correct this imbalance, a basement level of 1 million barrels per day (bpd) of exports should form the new basis for membership in OPEC. Eleven countries would qualify under such an arrangement, excluding current OPEC members Qatar, Indonesia, and Algeria, but also including Russia, Mexico, and Norway. Iraq would naturally be included, because its exports will continue to rise and it has the world's second largest proven reserves. Kazakstan, with proven reserves of around 20 billion barrels, could be another candidate for inclusion.

Small movements in the direction of OPEC reorganization have already appeared. In late 1998, Saudi oil minister Ali Al Naimi called for the creation of an "informal" petroleum group composed of the largest exporting countries. In addition, the primary accomplishment of OPEC meeting in Vienna in March 1999—Saudi production cuts—was negotiated outside traditional OPEC channels. Although Russia, Norway, and Mexico did not directly participate in these discussions, the Saudis and Iranians kept them closely informed of the proceedings.

If this trend continues, and if OPEC does not conjure up the will to reinvent itself, a new replacement organization of large-volume exporters may be created—perhaps called something like "OLPEC," the Organization of Large Petroleum Exporting Countries. Such an organization would control a larger share of world output and thus could wield much more influence on petroleum policy and pricing. A taste of this power was displayed in the Vienna cuts, as they were orchestrated by countries that would be the core of OLPEC, and they also had an immediate impact of raising oil prices. Of course, OLPEC advocates will have to deal with Norwegian,

Mexican, and Russian resistance to joining any organization similar to OPEC. Norway presents the biggest challenge because of the government's instinctive aversion to being part of a cartel. Mexico and Russia, both of which enjoy the benefits of "production management" (higher prices) without the same level of pain (lower production) that OPEC members suffer, would have reservations about joining. If the proponents could demonstrate that the long-term benefits outweigh the disadvantages, they may convince the three doubters to join this new group. The new members would benefit because the group could better manage prices and could avoid the dire economic situations they faced in the six months prior to March 1999.

A reorganization along these lines would make economic and political sense for the oil exporters. With 75 percent of world oil exports and 53 percent of total world oil output, this new group would have enormous power to manage oil production and prices in the short run—though the effectiveness of any cartel diminishes with time if it attempts to go beyond price stabilization by keeping prices substantially above what market forces would dictate. Finally, Mexico and Norway would share a more equitable burden of future production cuts, if they were necessary, to the advantage of current OPEC members. In the March 1999 cuts, Iran and Venezuela reduced their output by more than double the required cutback of the Mexicans and Norwegians, despite the comparable outputs of all four countries. Had Mexico and Norway been OPEC members at the time of the Vienna production agreement, perhaps an additional 250,000 bpd would have been taken off the market.

Such reorganization would not be easily accomplished, as it would also face severe resistance from those asked to leave. "We have no intention of leaving OPEC," one perturbed Qatari oil official said. "This whole idea is clearly a Saudi ploy undertaken for political rather than economic reasons."[2] Although the relationship between the kingdom and its neighbor Qatar have not been smooth recently, the economic figures alone are compelling enough to warrant Saudi Arabia making such a move without taking political motivations into account. Qatar and Algeria may balk at being asked to leave this exclusive club, but the higher petroleum prices that may result could ease the transition for them.

Oil Policy over the Long Term

Endless debate has occurred over the proper direction of Saudi Arabia's long-term oil policy. Should the kingdom work to maintain prices through global production management, or should it instead focus on increasing its market share, regardless of price? Many have advocated the latter approach, although they rightly point out that, in the short to medium term, the drop in oil prices would be difficult for the kingdom to bear. As the chart below shows, it would take five years for increased market share to compensate for decreased oil prices, a factor that is problematic. If this chasm can be bridged, however, it is clear that in the long term the Saudis would benefit.

The data used to create this chart assume several facts. For Option 1, the "price maintenance" strategy, continued

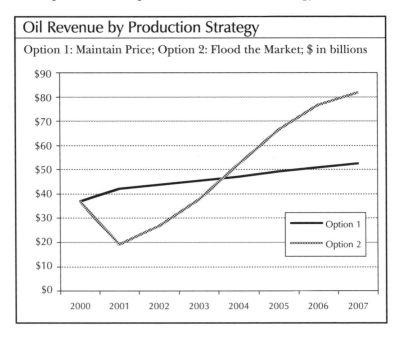

Oil Revenue by Production Strategy

Option 1: Maintain Price; Option 2: Flood the Market; $ in billions

cooperation within OPEC will hold Saudi crude at an average yearly price of $15 per barrel. Also, the kingdom's production cap will increase by 300,000 bpd from the March 1999 target of 7.4 million bpd.

Option 2, on the other hand, estimates what could occur if the kingdom began producing at capacity—a "flood-the-market" approach. It assumes the Saudi government would allow international oil companies (IOCs) to enter the upstream and downstream oil sectors to help the kingdom increase capacity to 16 million bpd. It also assumes that high-cost producers would quickly leave the market. The slope of Option 2 estimates a price drop to $5 per barrel in the first year with a slow recovery to $14 per barrel thereafter.

Under these assumptions, the second strategy would yield benefits for Saudi Arabia after seven years, when the kingdom would see a net gain of $30 billion ($400 billion vs. $370 billion total oil revenues). After this period, the Saudis would continue to earn $20 billion to $30 billion per year more than if they followed a price-maintenance approach. During the first five years, however, when increased market share has not yet offset low prices, the Saudi economy would suffer from decreased oil revenues—almost $50 billion in the third year. The experience of 1998, in which falling oil revenues wiped out about $20 billion from government revenues, provides ample evidence of the damage such a loss can cause.

To prevent severe economic dislocations and social unrest, the Saudis would have to raise additional funds to cover this expected $50 billion loss. One way to do so would be to move forward with the long-planned divestiture of state enterprises. A complete privatization of the four main state industries—Saudi Electric Company, Saudi Arabian Airlines, the Saudi Telecommunications Company, and the Saudi Arabian Basic Industries Corporation (SABIC)—could earn between $85 billion and $111 billion, including revenues from the sale of the companies themselves and returns from five years' worth of subsidies that the government would no longer have to pay. If the government continues its modest increases in taxes and fees, even a partial sale of the above industries

Estimated Returns on Sale of State Assets ($ billions)				
Entity	**Total Value**	**Privatization Windfall***	**100% Sale**	**50% Sell-Off**
Saudi Electric Company	15–20	11–13	26–33	13–17
Saudi Airlines	10–15	12–14	22–29	11–15
Telecom	12–17	5–7	17–24	8–12
SABIC	20–25	-	20–25	10–13
TOTAL	**57–77**	**28–34**	**85–111**	**42–57**

*"Privatization windfall" is the amount of money the government would save over five years by not paying subsidies for the entity in question.

Source: Author's personal calculations, based on classified data.

would more than cover the temporary losses incurred by producing oil at capacity. (In 1999 alone, through the higher price of gasoline for Saudi consumers, increased fees on foreign workers, and a departure fee at the airport, the government added $1 billion to its coffers.)

These figures only include the four major public enterprises. If Saudi Arabia were to sell all of its state-owned companies—the Saline Conversion Corporation, the General Organization for Technical and Vocational Training (GOTVOT), and the Jubail and Yanbu Utility Company—it could expect to reap an additional $30 billion to $40 billion dollars.[3] The Saudi government could avoid using these funds for other fiscal demands. Although the Saudi national debt is approximately $110 billion, ordinary revenue is adequate for those payments, and the Saudi Arabian Monetary Agency has enough reserves to maintain liquidity, cover emergency costs, and defend the currency.[4] Even if the Saudi government never chooses to implement such an oil policy, these calculations show that a partial sale of state assets could provide the cash necessary to tackle other structural and economic problems.

Aside from the economic benefits of having more than

$100 billion after ten years, the strategy outlined above has other advantages. It would free the Saudis from the need to constantly cajole and haggle with OPEC members. Untold Saudi political capital and human energy has been expended in the past thirty years to make OPEC work, with mixed results. That energy and talent could be directed toward solving some of the region's long-standing problems. Another benefit for Saudi Arabia would be that lower prices would put high-cost competitors such as Britain and Norway out of business for the long term and would spell the end of new oil development in locations such as Central Asia. Saudi Arabia would also benefit by slowing the ultimate adoption of emerging alternative energy technologies such as the electric car or the hydrogen fuel cell. The return of the IOCs would boost the Saudi economy, injecting competition into the energy sector, providing more jobs and training for Saudis, and helping the kingdom's case for membership in the World Trade Organization (WTO). Finally, and most important, the strategy outlined above would rationalize and stabilize Saudi Arabia's revenue stream. At present, Saudi revenues rely less on Riyadh's ability to provide the world with a high-quality product than on its ability to keep Caracas from cheating on the latest production quota. Of course, as long as the kingdom derives a majority of its revenue from oil, it will face the problems that all single-commodity economies must confront, but the removal of unstable supply management systems will eliminate one major source of volatility.

Notes

1. Author interview in Vienna, March 24, 1999.

2. Author interview in Vienna, March 24, 1999.

3. Personal calculations.

4. These were compiled from various sources, including individuals in the international banking industry, the intelligence services of G8 countries, and the finance ministries of several industrialized nations. The author is currently completing a more detailed analysis of this agency, which will reveal that the Saudi Arabian Monetery Agency has assets between $77 billion and $107 billion.

Senior Princes

After the king, the crown prince, and the minister of defense, the senior princes wield the most power in the kingdom. In fact, King Fahd has eleven full and half brothers in official governmental positions. In this section, an examination of their accomplishments in office will reveal their political and economic ideas. Although they do not currently exercise true influence over Saudi petroleum policy, in the

King Fahd's Full and Half Brothers in Official Positions		
Prince	**Year of Birth**	**Position**
Crown Prince Abdullah	1928	Vice Prime Minister, Commander of SANG
Sultan	1930	2nd Vice Prime Minister; Minister of Defense and Civil Aviation; Inspector General
Mitiab	1932	Minister of Public Works and Housing
Abdulrahman	1934	Vice Minister of Defense and Civil Aviation; Inspector General
Nayef	1935	Minister of Interior
Abdul Illah	1938	Governor of Jouf Province
Salman	1940	Governor of Riyadh Province
Ahmad	1945	Vice Minister of Interior
Abdulmajeed	1945	Governor of Makkah Province
Sattam	1947	Vice Governor of Riyadh Province
Miqrin	1948	Governor of Madinah Province

next decade or so they are expected to play a larger role in Saudi policymaking, especially the interior minister, the governor of Riyadh, the vice minister of the interior, and the director of general intelligence.

Prince Nayef, Minister of Interior

Prince Nayef bin Abdulaziz is the fourth full brother of the king. He began his public service as governor of Riyadh in 1953. He moved to the Interior Ministry, first serving as the vice minister under the future king, Fahd, and then as the minister of state for internal affairs. In 1975, when Prince Fahd was named crown prince, Prince Nayef was officially nominated as minister of interior, a position he has held since then. This position also holds the title of chairman of the Supreme Information Council, the governmental body that coordinates the collection and dissemination of information into the kingdom. Prince Nayef was the chairman of the By-laws Committee for the Consultative Council, in charge of drafting the laws of the Council; presided over the official creation of the Council in March 1992; and coordinated the appointment of the first 60 members in August 1993.[1]

As a former British Secret Intelligence Service station chief in Riyadh stated, "Prince Nayef created the best-funded domestic security system in all the Middle East and played a crucial role in maintaining the country's stability during the 1990s."[2] Prince Nayef's ministry comprises several security agencies, including the General Security Service (for counterintelligence and domestic security), the Public Security Agency (the police forces), the Border Guard Service, the Special Security Forces, the Anti-Drug Enforcement Agency, and the Civil Defense Administration.

After the 1979 hijacking of the Ka'aba in Mecca by Juhayman Al Otaibi and his group of followers, the kingdom realized it had to improve its security apparatus. Prince Nayef commanded the streamlining and modernizing of various security services. According to a former member of the U.S. Federal Bureau of Investigation (FBI) team that investigated the Al-Khobar bombing, "despite the fact that we did not always see eye to eye with Prince Nayef, I must admit that he has done a remarkable job of transforming the law enforcement agencies into an efficient and effective security force."[3] One important element was improving the professionalism of crime prevention and control. To this end, Prince Nayef established the King Fahd College for Security Sciences Studies and the Nayef Arab Academy for Security Studies to provide training in intelligence, security, and law enforcement for native Saudis and all citizens of the Arab League.

In addition, Prince Nayef was the prime architect behind the creation of the Gulf Cooperation Council (GCC) in 1981, established to increase security in an increasingly volatile region. In the years preceding its creation, the Iranian revolution and the beginning of the Iran–Iraq war sent shockwaves throughout the area. In 1980, the Soviet invasion of Afghanistan posed a third major political crisis in the region in the span of only two years, further increasing awareness of security issues. According to a former director of the Middle East department in the British Secret Intelligence Service (SIS), "Prince Nayef brought the Gulf States together, and convinced them of the necessity of forming security frameworks that ultimately developed into the GCC."[4] Although the GCC has not yet become an effective security and cooperation organization, it has provided a forum for the discussion of regional security issues and has allowed for more coordinated consultation with the United States.

One of the more noteworthy aspects of Prince Nayef's tenure is that under his command the security forces have maintained domestic tranquility while remaining more restrained, less repressive, and less feared that in any other large Middle Eastern state. "Saudi Arabia is a tightly closed society, but it tolerates a great deal of peaceful discussion and dissent."[5] The average Saudi respects but does not fear the state's security forces. According to a former FBI special agent, "Although the Interior Ministry has been able to keep law and order, sometimes through severe action, they do not have a history of repression."[6] On several occasions in recent years, the Interior Ministry handled dangerous situations with expertise and restraint. In the early 1990s, two vocal religious leaders—Shaykh Salman Al Awda and Shaykh Safar Al Hawali—denounced the government and called for sweeping reforms. The former distributed tapes of recorded sermons that compared members of the royal family to the last sultans of the Ottoman Empire and the Americans to an occupying force.[7] He claimed that the ruling family was "corrupt and incompetent" and ended by saying, "They do not deserve to rule if they don't practice Shari'a." After this act, the authorities decided to arrest them in 1994. His followers personally intervened and attempted to prevent the police from carrying out their orders. A videotape of the confrontation, distributed by the opposition group The Committee for the Defense of Legitimate Rights, reveals that the police acted with restraint. The Interior Ministry forces defused the potentially-explosive situation without injury and ultimately achieved the state's goals (Shaykh Al Awda was arrested). Another unreported demonstration occurred in Hail, and again the demonstrators and security forces achieved a peaceful denouement.[8]

Since then, they have arrested few dissenters, unlike the customary practices in GCC states like Bahrain or Oman, much less the draconian responses such protests would have drawn in Syria, Egypt, Iraq or Algeria. Although Interior Ministry forces arrested several hundred mid-level clergymen in the demonstrations described above, they have released nearly all of them. The Ministry kept Shaykh Al Awda under house arrest until June 1999, when they released him and two other clerics, Shaykh Al Hawali and Shaykh Nasser Al Omar.[9] A current high-ranking case officer in Pakistan's Inter-Services Intelligence Agency (ISI) made the point that the manner in which the senior princes handled the arrest, incarceration, and release of the three opposition figures reveals a great deal about the strength and restraint of the Saudi leadership compared to that of other governments in the region.

First of all, while the prisoners were under arrest they were provided with luxury accommodations and saw their fami-

lies every day. [Vice Minister of Interior Prince] Ahmad
made sure that they had proper medical attention (especially for Shaykh Al Awda, who had a kidney problem) and
that all their needs were met. Both the interior minister
and vice minister held meetings with them, where they tried
to understand their demands and points of view, and where
Prince Nayef expressed the government's perspective. And
now, not only have they been released, but [Prince] Nayef
has actually developed an ongoing line of communication
with them.[10]

Interior Ministry forces have dispersed smaller protests by
Shiites, women's rights advocates, and Islamists with no deaths
and few arrests. According to a former case officer of the GRU
(Soviet Military Intelligence) stationed in Marxist South Yemen
during most of the 1980s and 1990s, "While the Saudis don't
tolerate rebellious activity, they have shown that they do not
have to act repressively or with undue force to ensure social
stability and peace."[11] This approach seems to have worked
well. By the end of the decade, the domestic opposition movements had all but disappeared, largely because of the policies
of the Interior Ministry. "Nor is [fundamentalism] a challenge
to the regime as such—the Islamist threat has declined."[12]

Under Prince Nayef's supervision, the General Security
Service has overseen two major investigations of bomb attacks
on Saudi soil. In November 1995, a 200-pound car bomb killed
five U.S. military trainers and two Indians in Riyadh. The per-
petrators, four Saudi fundamentalists, were all apprehended
and tried and executed under Saudi law (like the United States,
Saudi Arabia has the death penalty for serious crimes). Nine
months after the first attack, a 5,000-pound bomb exploded
outside a U.S. military housing complex at Al-Khobar, killing
19 Americans and injuring 160 people. The Al-Khobar bombing remains unsolved, although Saudi security services have
identified the Saudi culprits. U.S. Attorney General Janet Reno
complained that "the kingdom did not cooperate as it should
have with the United States with respect to the Al-Khobar explosion investigations,"[13] but other senior officials have stressed

that this lack of cooperation has not unduly strained the relationship. "Of course we are upset and frustrated that there have been no arrests," Defense Department spokesman Ken Bacon said in June 1999. "However, we have pursued criminals in the past for years, even decades, and ultimately gotten them. We have strong relationships with Saudi Arabia."[14] A senior Clinton administration official said, "Frustration is one thing, but the strategic link with Saudi Arabia is very important to both of us in that region. We remain good and very solid friends."[15] Others have noted the constraints placed on Prince Nayef in his investigation. According to a case officer in Pakistan's ISI who was close to the talks, "it is true that Nayef was not always entirely candid with the Americans, but that was because he was balancing national security issues with the desire to solve this horrible crime."[16]

Prince Nayef has trained his sons in security issues. One of them, Prince Saud, is currently the deputy governor of the Eastern Province, and the second son, Prince Mohammed, was named in May 1999 as the assistant minister of interior for security affairs.

Prince Salman, Governor of Riyadh Province

Prince Salman is the sixth full brother of the king. As governor of Riyadh since 1962, he has supervised the rapid expansion of the city from a mid-sized town to a major urban metropolis larger than Paris and nearly equal to London in land area.[17] According to a former Saudi cabinet member, he is considered "a peacemaker and mediator who helps resolve internal governmental conflicts."[18] Former U.S. ambassador to Saudi Arabia Walter Cutler noted that "the governor is extremely popular with the general (native) populace of the kingdom."[19]

Prince Salman has always maintained a strong interest in petroleum matters, especially in Saudi relations with large consuming countries; his recent destinations have included the Philippines, Japan, and China. Saudi Aramco has sizable investments in the Philippines and Japan, and China will likely become a major importer of Saudi crude in the next decades.

During his April 1999 visit with Chinese president Ziang Zemin, he led the discussion on bilateral trade relations.[20] Prince Salman strongly supports increased international economic cooperation between the major industrialized countries and Saudi Arabia.

In addition to these concerns, Prince Salman is deeply involved in the Saudi foreign policy scene. According to Cutler, Prince Salman is "a staunch anti-Communist."[21] He has long advocated the need for strong ties to the West, especially to the United States. During the Cold War, he perceived the Soviet Union as the main threat to the basic principles governing the kingdom and was a major proponent of the Saudi effort to help defeat the Soviets in Afghanistan.[22]

Most important, his international affairs work has involved numerous relief efforts to aid Muslim nations in time of need, such as Somalia, the Sudan, Bangladesh, Afghanistan, and more recently Bosnia-Herzegovina. He was the chairman of the Saudi High Commission for Bosnia, which coordinated approximately $1.5 billion in aid to the former Yugoslav republic. In the words of the former International Committee of the Red Cross (ICRC) envoy to the Balkans, "the Saudis, along with the Americans, were the largest humanitarian donors to the Bosnian people."[23] During the Soviet occupation of Afghanistan, he was a principal architect of Saudi humanitarian and relief assistance to Afghanis stranded in Pakistani refugee camps. As one Pakistani ISI case officer stated, "Prince Salman's office provided billions of dollars to help us deal with the influx of Afghani refugees in the 1980s."[24]

Prince Salman has five sons of adult age. His eldest, Prince Fahd, the former vice governor of the Eastern Province, is currently a private entrepreneur based in Riyadh, mainly working in the real estate sector. His second son, Prince Sultan, a former officer in the Royal Saudi Air Force, flew on the U.S. space shuttle *Discovery* in 1985—the first Arab and the first Muslim astronaut. He is currently the chairman of the National Foundation for the Handicapped Children and is chairman of the Prince Salman Center for the Research on Handicapped Children. Prince Salman's third son, Prince

Ahmad, is the chief executive officer of the Arab world's largest media and publishing firm, the Saudi Research and Marketing Group. This company's flagship publication is *al-Sharq al-Awsat*, which possesses the largest circulation of any pan-Arab newspaper. Prince Abdulaziz, the fourth son of Prince Salman, is the current deputy minister of oil for petroleum affairs (see the section on the Petroleum Preparatory Committee). Finally, his fifth son, Prince Faisal, took a teaching position at the political science department of King Saud University in September 1999, after having graduated with a Ph.D. in international relations with a concentration in diplomatic history from St. Anthony's College at Oxford University. His dissertation was on Iranian foreign policy in the early 1970s under the shah.

Prince Ahmad, Vice Minister of Interior

Prince Ahmad is the seventh and last full brother of the king. He was the governor of the Mecca Region before becoming vice minister of interior in 1975. He works closely with Prince Nayef in the management of the Interior Ministry. Prince Ahmad holds a bachelor's degree in political science from Redlands University with a concentration in international security studies.

Prince Ahmad's accomplishments include assisting the interior minister in redirecting and overhauling the security apparatus of the kingdom in the 1980s. His main focus was operational developments and data-collecting techniques. He has directed the main organs of the Interior Ministry to combating the issues of terrorism, drug trafficking, and money laundering.

Prince Ahmad has also distinguished himself on regional security issues. According to a current officer in Pakistan's ISI, "Prince Ahmad was one of the main officials on the Saudi side to have helped defeat the Soviets in Afghanistan."[25] Also, a former GRU case officer posted in Aden, South Yemen, in the mid-1980s bluntly revealed that, "in implementing covert policy toward Saudi Arabia and the other Gulf states, we always perceived Prince Ahmad as one of seven officials—five

Saudis, one Bahraini, and one Omani—to analyze and watch closely. We held him in the highest esteem."[26] This focus on regional security issues also helped him obtain a leading role in the Riyadh bombing investigation. As a former FBI special agent said, "we were pleasantly surprised with the professionalism of Saudi law enforcement, but especially on two fronts: operations and collection." The same official noted that

> Prince Ahmad's technical knowledge of operational procedures helped us understand the peripheral problems that the Riyadh bombing entailed. . . . The coordination we got from the Ministry of Interior was not adequate at all, but Prince Ahmad always made a point to tell us why. Given the intense atmosphere after the bombing, our team found his candor extremely surprising.[27]

Prince Ahmad has two sons. The eldest, Prince Abdulaziz bin Ahmad, founded the Saudi Society for the Blind and Disabled, making it the most influential nonprofit organization to advocate and support the rights of the disabled in Saudi society. His other adult son, Prince Nayef, is a member of the Special Forces and currently a Ph.D. candidate in security studies at Cambridge University, focusing his research on national security issues facing the GCC states.

Notes

1. Hraire Dekmajian, "Saudi Arabia's Consultative Council," *Middle East Journal* (spring 1988), pp. 204–218.
2. Author interview in Washington, May 22, 1999.
3. Ibid.
4. Author interview in London, April 12, 1999.
5. Anthony H. Cordesman, *Saudi Arabia: Guarding the Desert Kingdom* (Boulder, Colo.: Westview, 1997), p. 176.
6. Author interview in Washington, May 22, 1999.
7. "Revolt in Buraiydah," released by the Committee for the Defense of Legitimate Rights, London, September 1993, videocassette.
8. Name withheld by request, interview with author in Paris, February 25, 1999 (former analysts in the Middle East Department of the Direction General de la Sécurité Exterieur (DGSE, or French intelligence service).

9. "Saudi Opposition Figures Freed, Group Says," Reuters, June 27, 1999.

10. Author interview in Baltimore, June 29, 1999.

11. Author interview in Bratislava, Slovakia, March 25, 1999.

12. David Hirst, "Fall of the House of Fahd," *Guardian* (London), August 11, 1999, p. 13.

13. "Paper Analyzes Sultan's Washington Visit," *Riyadh al-Jazirah*, March 4, 1997, p. 28 (translated from Arabic by the Foreign Broadcast Information Service (FBIS)).

14. Charles Aldinger, "Still No Charges Three Years After Saudi Blast," Reuters, June 24, 1999.

15. Ibid.

16. Author interview in Baltimore, June 29, 1999.

17. Douglas Jehl, "Riyadh Journal; Rival Princes Ease Desert City's Horizon Skyward," *New York Times*, April 6, 1999, p. A4.

18. Author interview in London, February 12, 1999.

19. Author interview in Washington, June 1, 1999.

20. "Chinese President Receives Saudi Prince Salman," Riyadh Saudi Arabian Television Network, April 23, 1999.

21. Author interview in Washington, June 1, 1999.

22. One analyst has argued that of $40 billion spent to fund the mujahedin, "the bulk [came] from the United States and Saudi Arabia, which contributed equally." Dilip Hiro, "The Cost of an Afghan 'Victory,'" *Nation*, January 30, 1999, p. 2–3.

23. Author interview in Geneva, March 12, 1999.

24. Author interview in Baltimore, June 29, 1999.

25. Author interview in Bratislava, Slovakia, March 25, 1999.

26. Author interview in Washington, January 16, 1999.

27. Author interview in Washington, June 29, 1999.

Appendix II
Prominent Third-Generation Princes

A mong the individuals who have an effect on Saudi
policymaking are the third-generation princes. Several
of them have successful careers in government service and,
through their present positions and qualifications, can be
expected to continue to play an ever-increasing role in Saudi
decision making.

Prince Saud Al Faisal, Ministry of Foreign Affairs

Prince Saud Al Faisal's background is detailed in chapter 3
in the section on the Supreme Council for Petroleum and
Minerals Affairs.

Prince Turki Al Faisal, Director of General Intelligence Directorate

Prince Turki Al Faisal graduated from Georgetown University's
School of Foreign Service with a bachelors of science in international security studies. In 1977, he assumed the
directorship of the kingdom's Foreign Intelligence Service
from his uncle, Kamal Adham. In this position, he closely
monitors all issues affecting Saudi national security, including the world energy market.

Under Prince Turki's direction, the Foreign Intelligence
Service has been modernized and streamlined. Moreover, he
has lobbied to include more civilians in the service, which
has been dependent on uniformed military officers. A Paki-
stani ISI case officer said, "We only have praise for Prince
Turki and the Saudi GID [as the Saudis are known in the
intelligence world]. They played an important role in not only
countering the communists in Afghanistan, but also in the
Middle East, especially in former South Yemen and Africa."[1]

A former lieutenant general in the Soviet Military Intelligence GRU corroborates the claim:

> The GID really gave us problems in the Middle East. Although in the 1980s we regarded the Saudi service as a second-tier organization well behind the Israelis, Egyptians, Iraqis, Jordanians and even the Syrians, the situation has drastically changed today. Prince Turki has really transformed the service.[2]

A former senior Central Intelligence Agency (CIA) case officer noted that "the Saudis, with huge resources and dire foreign security threats, have become in terms of collection and analysis, easily the second most-powerful intelligence service in the region after the Israelis." He emphasizes that "they fingered the Saudi Shi'a culprits belonging to Saudi Hizballah in the Al-Khobar bombing a month and a half after the horrific incident. This and other successes show that they have become a truly professional service, and most of the credit for that must go to [Prince] Turki."[3]

Prince Bandar bin Sultan, Ambassador to United States

Prince Bandar bin Sultan attended Johns Hopkins SAIS and graduated with a master's degree in international affairs. In 1983, he officially left the Royal Saudi Air Force, where he was an F-5 pilot, to become ambassador to the United States. Before that, he was the military attaché in charge of coordinating Saudi weapons purchases from the United States. He is currently the dean of the diplomatic corps in Washington, DC, that is, the longest serving ambassador.

Prince Bandar has been one of the primary architects of that "special relationship" that Saudi Arabia and the United States enjoy. He has helped promote U.S. interests not only in the Middle East but also in places like Central America, where he provided the initial funding to the Nicaraguan contras in their fight against the Marxist Sandinistas. He has also helped to maintain the Saudis as one of America's staunchest allies in the Middle East.[4] Prince Bandar initiated the idea of inviting major U.S. oil companies back into the kingdom, and to this end he orchestrated the meeting between the crown prince, the foreign ministers, and a handful of U.S. oil company CEOs in Washington in September 1998. It is not surprising that former U.S. ambassador to Saudi Arabia Walter Cutler proclaimed, "Prince Bandar is one of the most influential envoys in Washington today."[5]

Most recently, Prince Bandar's successful mediation in the Lockerbie affair (along with South African president Nelson Mandela) has consolidated his reputation as a "prolific" envoy, earning him popularity within the Saudi and the U.S. establishments. "I think Prince Bandar's success in Libya has really earned him respect and trust in both Washington and Riyadh," said a former assistant secretary of state.[6]

Prince Mitiab bin Abdullah, Vice Commander of Military Command, SANG

Prince Mitiab bin Abdullah is a graduate of Sandhurst Military Academy in England. His entire career has been spent in the Saudi Arabian National Guard (SANG). He established King Khaled Academy at the National Guard, which he currently heads. Prince Mitiab is extremely popular with SANG officers, especially with the large tribes of central Saudi Arabia who provide the bulk of the SANG forces. A former case officer in France's foreign intelligence service, DGSE (Direction General de la Securite Exterieur), noted this aspect when remarking, "Prince Mitiab's strength is not really government experience; it is the fact that he has a wide following in the kingdom's heartland."[7] A former British Special Air Services (SAS) officer who trained units in street-combat techniques says, "[Prince] Mitiab has, beyond any doubt, the full respect of the [SANG] troops, especially the ones I personally trained. I think he is the main advisor to his father [the crown prince], especially in military matters."[8]

Prince Abdulaziz bin Salman, Deputy Minister of Oil for Petroleum Policy

The prince's background is described in chapter 3, in the section on the ad hoc committees.

Prince Mohamad Bin Nayef, Assistant Interior Minister for Security Affairs

Prince Mohamad bin Nayef, son of the powerful interior minister, was appointed in May 1999 to his current position. Previously, he was an entrepreneur in the private sector concentrating on foreign joint ventures, especially with U.S. companies. According to a former British Special Air Services officer with wide experience in the kingdom, "Prince Mohamad is not new to the world of law enforcement and preventive security. Throughout his adult life he has been exposed and to some degree taught, through his father, the ins and outs of this very complex industry."[9] Although still in his mid-30s and having garnered a "strong professional reputation"[10] among Saudi businessmen, his background will undoubtly be one of his major assets as he consolidates his position as the number-three man in the Interior Ministry.

Prince Faisal Bin Salman, Assistant Professor of Political Science, King Saud University

Prince Faisal bin Salman, son of the Riyadh governor, was appointed to his current teaching position in September 1999. He received his Ph.D. in international relations from Oxford University's St. Anthony's College with a concentration in Middle Eastern diplomatic history. Although still in his first year of teaching, he seems already to have developed a following among his students. According to a second-year student who had Prince Faisal as a teacher, he and his friends "were amazed to have Professor Faisal teach us the way he did. I mean, he insisted [that we] argue points he made that we did not agree with, discuss sensitive political issues, and stop him in the middle of a lecture if we really had a pressing question that needed clarifications."[11]

Prince Abdulaziz bin Fahd, Chief of Council of Ministers and Minister of State

Prince Abdulaziz bin Fahd was appointed to his current government job in January 2000. Prior to that, he had been a minister of state without portfolio since 1998. He graduated from King Saud University with a bachelor's degree in political science. Because of his relative youth—he is in his late 20s—he does not have the international exposure and administrative experience of his cousins. Nevertheless, Prince Abdulaziz has emerged as one of King Fahd's key aides in the last five years. According to a former U.S. ambassador to Saudi Arabia, "Prince Abdulaziz—through being always by the side of his father, the King—has gained enormous training and insights on how the vast Saudi administration functions. In that capacity, he has clearly become an asset to his uncles as they are slowly beginning to install much needed economic and bureaucratic reforms."[12]

Notes

1. Author interview in Baltimore, June 9, 1999.
2. Author interview in Bratislava, Slovakia, March 25, 1999.
3. Author interview in Washington, July 16, 1999.
4. Bob Woodward, *Veil* (New York: Simon and Schuster, 1994).
5. Author interview in Washington, June 29, 1999.
6. Author interview in Washington, July 25, 1999.
7. Author interview in Paris, February 25, 1999.
8. Author interview in London, April 20, 1999.
9. Author interview in Riyadh, December 28, 1999.
10. Author interview with four prominent Saudi businessmen in Al Kharj, January 5, 2000.
11. Author interview in Riyadh, December 30, 1999.
12. Author interview in Riyadh, December 18, 1999.

Charts and Graphs

Chart 1: OPEC Spare Production Capacity

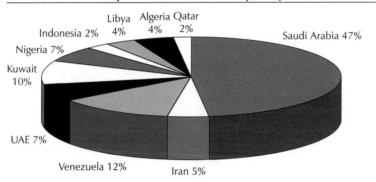

Chart 2: Brent Prices, April 1998–July 1999

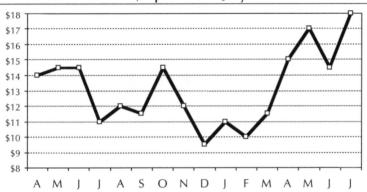

Chart 3: Saudi Current Account ($ in millions)

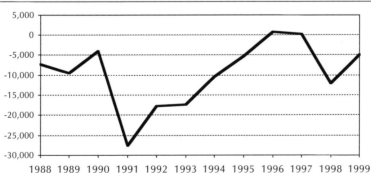

Chart 4: U.S. Oil Imports (millions of barrels per day)

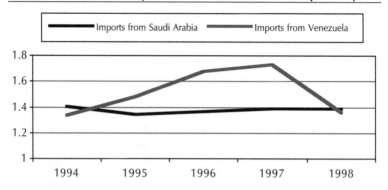

Chart 5: U.S. Arms Exports to Saudi Arabia ($ billions)

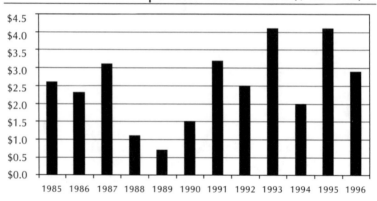

Chart 6: Major Oil Exporters (as of March 1999)

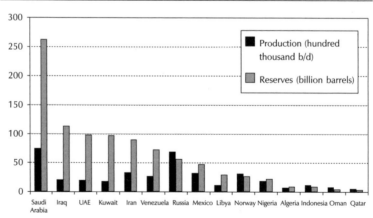

Chart A: Power Generation Capacity

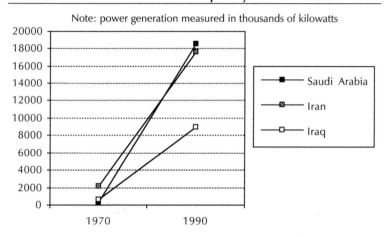

Note: power generation measured in thousands of kilowatts

Saudi Arabia
Iran
Iraq

Chart B: Adult Literacy (Percentage of Population)

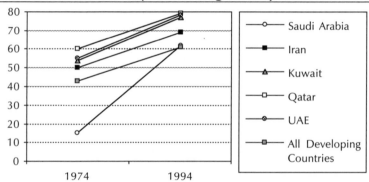

Saudi Arabia
Iran
Kuwait
Qatar
UAE
All Developing Countries

Chart C: Life Expectancy (Years)

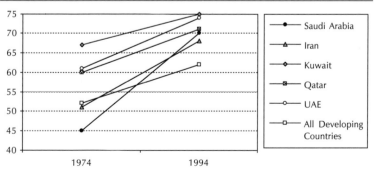

Saudi Arabia
Iran
Kuwait
Qatar
UAE
All Developing Countries

Chart D: Main Telephone Lines (in thousands)

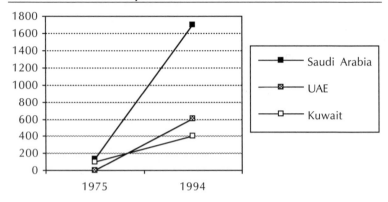

Chart E: Population per Physician

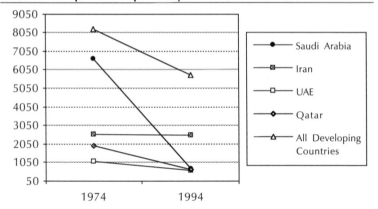

Chart F: Access to Safe Water (% of population)

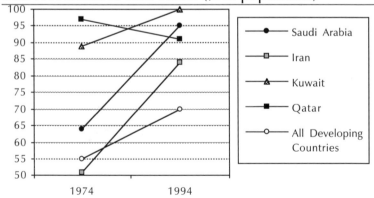

The Washington Institute's
Recent Publications

Palestinian Democracy & Governance: An Appraisal of the Legislative Council

David Schenker

An in-depth assessment of thr Palestinian Legislative Council's role in the peace process and its power in relation to Yasir Arafat's executive authority in the Palestinian Authority. Considers what the PLC's governing experiences and growing pains portend for the growth of democracy in a future Palestinian state. *Policy Paper no. 51 (2000) ISBN 0-944029-34-5 $19.95*

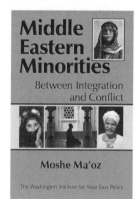

Middle Eastern Minorities: Between Integration and Conflict

Moshe Ma'oz

A broad survey of the historic and current role of religious and ethnic minorities in the Fertile Crescent, Egypt, and Sudan. Focuses on Shi'i–Sunni, Jewish–Muslim, Christian–Muslim, Kurdish–Arab, and similar relations from the mid-1800s to the present, and suggests ways that Washington policymakers can promote the political, cultural, and religious rights of minorities. *Policy Paper no. 50 (1999) ISBN 0-944029-33-7 $19.95*

Crises after the Storm: An Appraisal of U.S. Air Operations in Iraq since the Persian Gulf War

Lt. Col. Paul K. White, U.S. Air Force

The second book in The Washington Institute's new series of Military Research Papers written by visiting military fellows, *Crises after the Storm* analyzes the U.S. Air Force's objectives in containing Saddam Husayn's regime in Iraq since 1991. Reviews the four main crises that have involved large-scale troop deployments to the Gulf region this decade and highlights some of the key "lessons learned" from these crises. *MRP no. 2 (1999) ISBN 0-944029-32-9 $19.95*

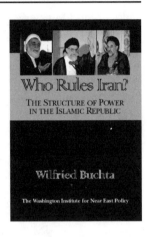
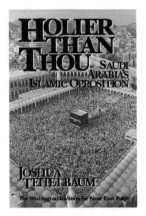